FORTRAN PROGRAMS
for Scientists
and
Engineers

FORTRAN PROGRAMS
for Scientists
and
Engineers

Alan R. Miller

Berkeley • Paris • Düsseldorf

Cover Design by Daniel Le Noury
Technical illustrations by j. trujillo smith, Jeanne E. Tennant
Book layout and design by Ingrid Owen.

FORTRAN-80 is a registered trademark of Microsoft Consumer Products.
DEC-20 is a registered trademark of Digital Equipment Corporation.
Lifeboat 2.2 is a registered trademark of Lifeboat Associates.
WordStar is a registered trademark of MicroPro International Corporation.
Z80 is a registered trademark of Zilog, Inc.

SYBEX is not affiliated with any manufacturer.

Library of Congress Card Number: 82-80263
ISBN 089588-082-2
First Edition 1982
Printed in the United States of America
10 9 8 7 6 5 4

Contents

Preface *xi*

Introduction *xiii*

A Note on Typography *xvii*

1 *Evaluation of a FORTRAN Compiler* 1

Precision and Range of Floating-Point Operations   1
FORTRAN Program: A Test of the Floating-Point
    Operations   2
FORTRAN SIN and COS Functions   4
FORTRAN Program: Testing the SIN Function   4
FORTRAN Program: Double Precision Test   7
FORTRAN Program: Arguments to Built-in Functions   8
FORTRAN Program: Entering Data from the Keyboard   9
Summary   10
Exercises   11

2 *Mean and Standard Deviation* 13

The Mean   13
The Standard Deviation   15
FORTRAN Program: Mean and Standard Deviation   17
Random Numbers   20
FORTRAN Function: A Random Number Generator   20
FORTRAN Program: Evaluation of a Random Number
    Generator   21

FORTRAN Program: Generating and Testing Gaussian
   Random Numbers   24
Summary   26
Exercises   27

3   *Vector and Matrix Operations*   **29**

Scalars and Arrays   29
Vectors   30
Matrices   33
FORTRAN Program: Matrix Multiplication   37
Determinants   40
FORTRAN Program: Determinants   41
Inverse Matrices and Matrix Division   43
Summary   43
Exercises   44

4   *Simultaneous Solution of Linear Equations*   **47**

Linear Equations and Simultaneous Solutions   47
Solution by Cramer's Rule   49
FORTRAN Program: A More Elegant Use of Cramer's
   Rule   52
Solution by Gauss Elimination   55
FORTRAN Program: The Gauss Elimination Method   58
Solution by Gauss-Jordan Elimination   63
FORTRAN Program: Gauss-Jordan Elimination   64
Multiple Constant Vectors and Matrix Inversion   69
FORTRAN Program: Gauss-Jordan Elimination with Multiple
   Constant Vectors   71
Ill-Conditioned Equations   74
FORTRAN Program: Solving Hilbert Matrices   76
A Simultaneous Best Fit   78
FORTRAN Program: The Best-Fit Solution   79
Equations With Complex Coefficients   82
FORTRAN Program: Simultaneous Equations with Complex
   Coefficients   85
The Gauss-Seidel Iterative Method   89
Summary   95
Exercises   96

5 *Development of a Curve-Fitting Program* **99**

The Main Program   100
A Printer Plotter Routine   103
A Simulated Curve Fit   108
The Curve-Fitting Algorithm   111
The Correlation Coefficient   116
FORTRAN Program: Least-Squares Curve-Fitting for Simulated
   Data   118
Summary   121
Exercises   122

6 *Sorting* **125**

Handling Experimental Data   125
A Bubble Sort   127
FORTRAN Program: The Bubble Sort   127
FORTRAN Program: Bubble Sort with SWAP   130
A Shell Sort   133
FORTRAN Program: The Shell-Metzner Sort   133
The Quick Sort   134
FORTRAN Program: A Nonrecursive Quick Sort   134
Incorporating SORT into the Curve-Fitting Program   136
Summary   137
Exercises   137

7 *General Least-Squares Curve Fitting* **139**

A Parabolic Curve Fit   140
FORTRAN Program: Least-Squares Curve Fit for a
   Parabola   141
Curve Fits for Other Equations   144
FORTRAN Program: The Matrix Approach to Curve
   Fitting   147
FORTRAN Program: Adjusting the Order of the
   Polynomial   152
FORTRAN Program: The Heat-Capacity Equation   156
FORTRAN Program: The Vapor Pressure Equation   160
A Three-Variable Equation   163
FORTRAN Program: An Equation of State for Steam   164

Summary   170
Exercises   171

8 Solution of Equations by Newton's Method   175

Formulating Newton's Method   176
FORTRAN Program: A First Attempt at Newton's
    Method   181
FORTRAN Program: Solving Two Different Equations   190
FORTRAN Program: Solving Other Equations   191
FORTRAN Program: The Vapor Pressure Equation   194
Summary   195
Exercises   196

9 Numerical Integration   199

The Definite Integral   200
The Trapezoidal Rule   201
FORTRAN Program: The Trapezoidal Rule with User Input for
    the Number of Panels   203
FORTRAN Program: An Improved Trapezoidal Rule   204
FORTRAN Program: Trapezoidal Rule with End
    Correction   207
FORTRAN Program: Simpson's Integration Method   209
FORTRAN Program: The Simpson Method with End
    Correction   213
The Romberg Method   215
FORTRAN Program: Integration by the Romberg
    Method   216
Functions that Become Infinite at One Limit   220
FORTRAN Program: Adjustable Panels for an Infinite
    Function   220
Summary   222
Exercises   223

10 Nonlinear Curve-Fitting Equations   225

Linearizing the Rational Function   226
FORTRAN Program: The Clausing Factor Fitted to the Rational

Function   226
Linearizing the Exponential Equation   230
FORTRAN Program: An Exponential Curve Fit for the Diffusion
    of Zinc in Copper   230
Direct Solution of the Exponential Equation   234
FORTRAN Program: A Nonlinearized Exponential
    Curve Fit   236
Summary   240
Exercises   240

11 Advanced Applications:
The Normal Curve, the Gaussian Error Function, the Gamma Function, and the Bessel Function   243

The Normal and Cumulative Distribution Functions   244
The Gaussian Error Function   246
FORTRAN Program: Evaluating the Gaussian Error Function
    Using Simpson's Rule   248
FORTRAN Program: Evaluating the Gaussian Error Function,
    Using an Infinite Series Expansion   250
The Complement of the Error Function   252
FORTRAN Program: Evaluating the Complement of the Error
    Function   253
The Gamma Function   255
FORTRAN Program: Evaluation of the Gamma
    Function   257
Bessel Functions   260
FORTRAN Program: Bessel Functions of the First Kind   261
FORTRAN Program: Bessel Functions of the Second
    Kind   263
Summary   266
Exercises   267

Appendix A: Reserved Words and Functions   269

Some Reserved Words   269
Some FORTRAN Functions   269
Format Descriptors   270
Carriage Control   270

Appendix B: Summary of FORTRAN   271

The FORTRAN Character Set   271
Variable Names   271
Array Variables   272
Data Statements   272
Constants   272
Comments   273
Operations   273
Assignment Statements   274
The Unconditional Branch   274
Conditional Branching   274
Iterative Statements   275
Input and Output   275
Subroutines   276

Bibliography   277
Index   279

Preface

The ideas and material for this book have been developed during my experience teaching numerical methods to sophomore, junior, and senior engineering students over the past 15 years. I have used FORTRAN, BASIC, and Pascal as the computer languages for these courses.

All of the FORTRAN programs in this book were developed on a Z80 microcomputer. The operating system was the Lifeboat 2.2 version of CP/M. The source programs were written with MicroPro's WordStar and compiled with Microsoft's FORTRAN-80, Version 3.4. Most of the programs were also run on a DEC-20.

The manuscript was created and edited with MicroPro's WordStar running on the same Z80 computer. The FORTRAN source programs have been incorporated directly into the manuscript from the original source files. Computer printouts shown in the figures were also incorporated magnetically into the manuscript. This was accomplished by altering the CP/M operating system so that printer output was written into a block of memory. This block was then saved as a disk file. The final manuscript was submitted to SYBEX in a magnetic form compatible with the photocomposer. Consequently, the manuscript and the FORTRAN source programs have not been retyped.

Alan R. Miller
Socorro, New Mexico
February 1982

Introduction

The purpose of this book is twofold: to help the reader develop a proficiency in the use of the FORTRAN language, and to build a library of programs that can be used to solve problems frequently encountered in science and engineering.

The programs in this book will prove valuable to the practicing scientist or engineer. The material is also suitable for a junior- or senior-level engineering course in numerical methods. The reader should have a working knowledge of an applications language such as FORTRAN, BASIC or Pascal. In addition, experience with vector operations and with differential and integral calculus will be helpful.

FORTRAN is one of the oldest computer languages. Consequently, it is usually implemented on all large and medium computers. FORTRAN is less commonly found on microcomputers. Nevertheless, several versions of FORTRAN are available for micros.

FORTRAN has evolved from FORTRAN II to FORTRAN IV (more recently known as FORTRAN 66) to FORTRAN 77. FORTRAN 77, which incorporates block structure, is beginning to appear on larger computers, but it is not yet available on microcomputers. As a consequence, since some readers may only have access to microcomputers, the programs in this book are written in FORTRAN 66.

The FORTRAN language generally requires logical unit numbers for performing input and output to physical devices such as the console and printer. However, there is no standard system for assigning such numbers. Therefore, throughout this book, the symbols IN and OUT are used for console input and output. For example:

```
    READ(IN, 101) LIST
```
and
```
    WRITE(OUT, 102) LIST
```

The symbols IN and OUT are declared in a COMMON block which is given the name INOUT. Values are assigned to IN and OUT near the beginning of each main program. Any subroutine that performs input or output can obtain the required logical unit number from the common block. For the examples in this book, IN and OUT are defined as unity to be compatible with the Microsoft compiler. You may need to change the values to something else for your particular FORTRAN.

The reader who is primarily interested in the FORTRAN programs developed in this book will have no trouble locating them; the sections that contain programs or subroutines are clearly labeled, and each program listing or output is displayed as a figure. However, this book is designed to be read from beginning to end. Each chapter discusses and develops tools that will be used again in subsequent chapters. The mathematical algorithms of each program are methodically described before the program itself is implemented, and sample output is supplied for most of the programs. The following brief descriptions summarize the contents of each chapter.

Chapter 1, *Evaluation of a FORTRAN Compiler*, identifies weak points in several commercial FORTRANs, and supplies programs for testing any FORTRAN. The results will be used to select various constants and operations in later chapters.

Chapter 2, *Mean and Standard Deviation*, discusses some basic statistical algorithms and presents a program for implementing them. Routines for generating—and testing—both uniform and Gaussian random numbers are also given.

Chapter 3, *Vector and Matrix Operations*, summarizes the operations of vector and matrix arithmetic, including dot product, cross product, matrix multiplication and matrix inversion. Two important programs are developed—one for carrying out matrix multiplication, and another for calculating determinants.

Chapter 4, *Simultaneous Solution of Linear Equations*, presents programs to carry out the algorithms of Cramer's rule, the Gauss elimination method, the Gauss-Jordan elimination method, and the Gauss-Seidel method—all for solving simultaneous equations. In addition, ill-conditioning is studied by observing a program that generates Hilbert matrices, and a program is developed for solving equations with complex coefficients.

Chapter 5, *Development of a Curve-Fitting Program*, is the first of a series of chapters on curve fitting. In a good illustration of top-down program development, a linear least-squares curve-fitting program is written and discussed. The program includes routines to simulate data, plot curves, compute the fitted curve, and supply the correlation coefficient.

Chapter 6, *Sorting*, describes and compares several FORTRAN sorting routines including two bubble sorts, a Shell sort and a nonrecursive quick sort. A sorting routine is incorporated into the curve-fitting program of Chapter 5 to enable the program to handle real experimental data.

Chapter 7, *General Least-Squares Curve Fitting*, extends the curve-fitting program to general polynomial equations, and finds curve fits for three examples: heat capacity, vapor pressure and the properties of superheated steam.

Chapter 8, *Solution of Equations by Newton's Method*, presents a series of programs that use Newton's algorithm for finding the roots of an equation. This tool will be used again in Chapter 10 for nonlinear curve fitting.

Chapter 9, *Numerical Integration*, develops programs for three different integration methods—trapezoidal rule, Simpson's rule, and the Romberg method. End correction is also discussed. Simpson's rule will be used in Chapter 11 for evaluating the Gaussian error function.

Chapter 10, *Nonlinear Curve Fitting Equations*, discusses curve-fitting algorithms for the rational function and the exponential function. Two examples are given—the Clausing factor, and the diffusion equation.

Chapter 11, *Advanced Applications: The Normal Curve, the Gaussian Error Function, the Gamma Function and the Bessel Functions*, addresses several advanced topics in programming for mathematical applications. This last chapter summarizes and expands upon a number of the concepts presented earlier in the book.

Each chapter also contains exercises designed to extend the reader's comprehension of the material.

For the reader who is approaching FORTRAN for the first time, a summary of the syntax, standard functions, and reserved words of FORTRAN is included in the appendices. The real educational experience of this book, however, will be gained by carefully working through the programs themselves.

A Note on Typography

The manuscript of this book was submitted to SYBEX on magnetic disks and transcribed by our in-house computers. The program listings and their accompanying outputs have been removed from the text and photographed directly from a daisy-wheel printer. At no point has the author's source code been retyped.

Those program lines reproduced in the text have been set in the typeface know as Futura, which resembles printer output. The text itself has been set in Oracle type.

Throughout this book an effort has been made to distinguish typographically between program structures and mathematical values. FORTRAN reserved words, variable names, operators, etc., have been capitalized. Mathematical expressions (variables and letter constants) appear in *italics*, except for vectors, which appears in **boldface** roman type. For example:

$$A + Bx + Cx^2 = 0$$
$$\mathbf{v} = [1\ 2\ 3]$$

Juxtaposed in a single paragraph, therefore, the reader may see references to the variable x and the FORTRAN variable X; the vector $\mathbf{v}$ and the FORTRAN array V; the matrix element v_{ij} and the FORTRAN array element V(I,J).

1

Evaluation of a FORTRAN Compiler

To UNDERSTAND THE RESULTS of a FORTRAN program we must be familiar with the limitations of the compiler we are using. This is particularly true of scientific-application programs such as those given in this book. In this first chapter, then, we will present some tools for evaluating the precision and range of any FORTRAN compiler. In fact, the examples given here were derived from three commercially available FORTRAN compilers.

PRECISION AND RANGE OF FLOATING-POINT OPERATIONS

Many of the programs in this book are sensitive to the *precision* and *dynamic range* of the FORTRAN floating-point operations. For example, in one program an algorithm is terminated when a particular term is smaller than a relative tolerance. The formula in this case is:

$$TERM < TOL * SUM$$

where TERM is the value of the new term, SUM is the current total, and TOL is an arbitrarily small number known as the *tolerance*.

It is important that the value chosen for the tolerance not be outside the accuracy of the floating-point operations. Otherwise, the summation step will never terminate. Suppose, for example, that the floating-point operations are performed to a precision of six significant figures. Then, the tolerance must be set to a value larger than 10^{-6}.

The dynamic range of the exponent is a separate matter. Typical binary, floating-point operations are performed with 32 bits of precision. BCD floating-point packages, on the other hand, will usually have a greater dynamic range.

We will now present a FORTRAN program for testing the precision and dynamic range of a compiler. We will investigate output from several compilers to illustrate both mantissa and exponent accuracy.

FORTRAN PROGRAM: A TEST OF THE FLOATING-POINT OPERATIONS

The program given in Figure 1.1 can be used to determine the precision and the dynamic range of a FORTRAN compiler. Type up the program and execute it. You may have to change the value of OUT to the corresponding logical unit number for your FORTRAN. The initial value of X is obtained by dividing 10^{-4} by 3. Then, successively smaller and smaller values of X are calculated and displayed on the console. Each succeeding value is obtained by dividing the previous value by 10. The process continues until 75 values have been printed or until a floating-point error terminates the program.

```
C       PROGRAM TEST
C
C -- Test significance and range of the
C -- floating-point operations.
C
        REAL X
        INTEGER I, N, OUT
C
        OUT = 1
        N = 75
        X = 1.0E-4 / 3.0
        DO 10 I = 1, N
          X = X / 10.0
          WRITE(OUT, 101) I, X
10      CONTINUE
        STOP
101     FORMAT(1X, I4, 1PE15.6)
        END
```

Figure 1.1: A Test of the Floating-Point Operations

Let us now use this program to test the accuracy and range of two different compilers.

Two Runs of the Program: A Comparison

The initial mantissa is chosen to be 1/3, a repeating fraction that cannot be precisely represented by a floating-point number. Successive multiplications will show the extent of roundoff error. A 32-bit binary, floating-point number typically utilizes three bytes for the mantissa and one byte for the exponent. This usually produces six or seven significant figures of precision and a dynamic range of 10^{+39} to 10^{-39}. The result might look like the output in Figure 1.2. In this example, the dynamic range goes to 10^{-39} and accuracy of the mantissa is about seven significant figures. When floating-point underflow occurs, an error message is displayed and a value of zero is substituted for the result.

```
 1    3.333333E-06
 2    3.333333E-07
 3    3.333333E-08
 4    3.333333E-09
 5    3.333333E-10
 6    3.333333E-11
 7    3.333333E-12
 8    3.333333E-13
 9    3.333333E-14
10    3.333333E-15
 .  .  .
 .  .  .
30    3.333333E-35
31    3.333333E-36
32    3.333333E-37
33    3.333333E-38
34    3.333333E-39
      (Floating-point error message)
35    0.000000E+00
36    0.000000E+00
37    0.000000E+00
38    0.000000E+00
39    0.000000E+00
40    0.000000E+00
```

Figure 1.2: Precision Test: Output from First Compiler

As another example, consider the output (shown in Figure 1.3) from a different FORTRAN compiler that also uses 32-bit binary floating-point operations. The results in this case are similar to the previous example except that the mantissa is a little less accurate. Notice, however, that the exponential range is greater, going to 10^{-76}.

```
      1     3.33333E  -6
      2     3.33333E  -7
      3     3.33333E  -8
      4     3.33333E  -9
      5     3.33333E -10
      6     3.33333E -11
      7     3.33333E -12
      8     3.33333E -13
      9     3.33333E -14
     10     3.33333E -15
      .  .  .
      .  .  .
     61     3.33333E -66
     62     3.33333E -67
     63     3.33333E -68
     64     3.33332E -69
     65     3.33332E -70
     66     3.33333E -71
     67     3.33333E -72
     68     3.33333E -73
     69     3.33333E -74
     70     3.33332E -75
     71     3.33333E -76
     72     (Error message from compiler)
```

Figure 1.3: Precision Test: Output from Second Compiler

A rather interesting bug contained in some FORTRAN compilers can limit the range of the SIN and COS functions. We will now investigate this phenomenon.

FORTRAN SIN AND COS FUNCTIONS

A problem can occur with the SIN and COS functions as the argument approaches zero. When the magnitude of the argument is less than 10^{-8} or so, the SIN function should return the argument and the COS function should return unity. But some commercial FORTRAN compilers show significant roundoff error. The argument is squared before a range check is performed. This produces a floating-point underflow and meaningless results.

In the following section we will present a program for studying this problem and test three compilers with the program.

FORTRAN PROGRAM: TESTING THE SIN FUNCTION

The program given in Figure 1.4 can be used to check the SIN function of your FORTRAN. Type up the program and execute it. Be sure that the value of OUT is properly set to the logical unit number for your output.

```
C        PROGRAM STEST
C
C -- Test significance and range of SIN function.
C
         REAL X, S
         INTEGER I, N, OUT
C
         OUT = 1
         N = 75
         X = 1.0E-4 / 3.0
         DO 10 I = 1, N
           X = X / 10.0
           S = SIN(X)
           WRITE(OUT, 101) I, X, S
10       CONTINUE
         STOP
101      FORMAT(1X, I4, 1P2E15.6)
         END
```

Figure 1.4: Test for the SIN Function

Running the SIN Test: Three Compilers

If your built-in SIN function correctly handles small numbers, meaningful values should be returned over the entire dynamic range of the floating-point operations. In this case, floating-point underflow should occur at the same place as for the previous test. In the example shown in Figure 1.5, the dynamic range of the floating-point operations (the second column) and the limit of the SIN function (the third column) are both 3.3E−39.

Some FORTRAN implementations of the SIN function initially square

```
 1    3.333333E-06    3.333333E-06
 2    3.333333E-07    3.333333E-07
 3    3.333333E-08    3.333333E-08
 4    3.333333E-09    3.333333E-09
 5    3.333333E-10    3.333333E-10
 6    3.333333E-11    3.333333E-11
 7    3.333333E-12    3.333333E-12
 8    3.333333E-13    3.333333E-13
 9    3.333333E-14    3.333333E-14
10    3.333333E-15    3.333333E-15

   . . .

   . . .
30    3.333333E-35    3.333333E-35
31    3.333333E-36    3.333333E-36
32    3.333333E-37    3.333333E-37
33    3.333333E-38    3.333333E-38
34    3.333333E-39    3.333333E-39
      (Error message from compiler)
```

Figure 1.5: SIN test: Output from First Compiler

the argument without performing a range check. In this case, a floating-point underflow will occur much too soon. For the second compiler, shown in Figure 1.6, the normal dynamic range is 10^{+76} to 10^{-76}. But floating-point underflow occurs during calculations of the SIN when the argument is less than 10^{-37}.

1	3.33333E -6	3.33333E -6
2	3.33333E -7	3.33333E -7
3	3.33333E -8	3.33333E -8
4	3.33333E -9	3.33333E -9
5	3.33333E-10	3.33332E-10
6	3.33333E-11	3.33333E-11
7	3.33333E-12	3.33333E-12
8	3.33333E-13	3.33333E-13
9	3.33333E-14	3.33333E-14
10	3.33333E-15	3.33333E-15
11	3.33333E-16	3.33333E-16
12	3.33333E-17	3.33333E-17
13	3.33333E-18	3.33333E-18
14	3.33333E-19	3.33333E-19
15	3.33333E-20	3.33332E-20
. . .		
. . .		
28	3.33332E-33	3.33332E-33
29	3.33332E-34	3.33332E-34
30	3.33333E-35	3.33332E-35
31	3.33333E-36	3.33333E-36
32	3.33332E-37	3.33332E-37
(Error message from compiler)		

Figure 1.6: SIN Test: Output from Second Compiler

In the next example, shown in Figure 1.7, the dynamic range of the floating-point operations is 10^{+39} to 10^{-39}. However, in this case the SIN function returns a value of 10^{-38} for arguments between 10^{-7} and 10^{-38}.

1	3.333333E-06	3.370563E-06
2	3.333333E-07	3.745070E-07
3	3.333333E-08	9.232311E-39
4	3.333333E-09	9.232311E-39
5	3.333333E-10	9.232311E-39
6	3.333333E-11	9.232311E-39
. . .		
. . .		

Figure 1.7: SIN Test: Output from Third Compiler

If you find that the SIN function is incorrectly implemented in your FORTRAN, you may have to use it with care. The subroutine given in Figure 8.16 shows how this can be done.

FORTRAN PROGRAM: DOUBLE PRECISION TEST

Standard FORTRAN includes a double precision floating-point representation, which more than doubles the number of significant figures. The dynamic range, however, is usually the same as for the single precision representation. The program given in Figure 1.8 can be used to investigate the double precision representation. Type up this program and execute it. Compare the results to Figure 1.9.

```
C       PROGRAM DTEST
C
C -- Test significance and range of double-precision
C -- floating-point operations.
C
        DOUBLEPRECISION X
        INTEGER I, N, OUT
C
        OUT = 1
        N = 75
        X = 1.0D-4 / 3.0
        DO 10 I = 1, N
          X = X / 10.0
          WRITE(OUT, 101) I, X
10      CONTINUE
        STOP
101     FORMAT(1X, I4, 1PD23.15)
        END
```

Figure 1.8: Test for DOUBLE PRECISION Operations

```
 1    3.333333333333333D-06
 2    3.333333333333333D-07
 3    3.333333333333333D-08
 4    3.333333333333333D-09
 5    3.333333333333333D-10
 6    3.333333333333333D-11
 7    3.333333333333333D-12
 8    3.333333333333333D-13
 9    3.333333333333333D-14
10    3.333333333333333D-15
 . . .
 . . .
30    3.333333333333333D-35
31    3.333333333333333D-36
32    3.333333333333333D-37
33    3.333333333333333D-38
34    0.000000000000000D+00
35    0.000000000000000D+00
```

Figure 1.9: DOUBLE PRECISION: Output from Test

FORTRAN PROGRAM: ARGUMENTS TO BUILT-IN FUNCTIONS

Constants, variables and functions in FORTRAN programs are typed as REAL, INTEGER, DOUBLE PRECISION, or LOGICAL entities. Care must be taken to avoid the improper mixing of types. If RE is a real variable and IN is an integer variable, then the FORTRAN expression:

RE = IN

properly sets the value of RE to the value of IN. Furthermore, mixed-mode operations are usually allowed in mathematical expressions. For example, the FORTRAN expression:

RE = 2.5 + 10

sets the value of RE to the sum of a real number and an integer.

The arguments to subroutines and functions are a special case. The type of the argument must always match the declared type. For example, the argument to the square root function, SQRT, must be a real number. Similarly, the argument to the double precision square root function, DSQRT, must be a double precision number. If the type of an argument is incorrect, the result will likely be in error. Some FORTRAN compilers will identify errors of this nature. Others, however, will simply give the wrong answer.

The program given in Figure 1.10 can be used to investigate this potential problem. The program attempts to find the square root of 2 in three ways. The first two methods are correct because the argument is a real number. The third method, however, is improper because the argument is an integer.

```
C       PROGRAM TSQRT
C
C  --  Test argument of the SQRT function.
C
        REAL X, SX, SR, SI
C
        OUT = 1
        X = 2.0
        SX = SQRT(X)
        SR = SQRT(2.0)
        SI = SQRT(2)
        WRITE(1,101) SX, SR, SI
        STOP
101     FORMAT(//' The square root of 2'/ 1X, 3F10.5)
        END
```

Figure 1.10: Testing the Argument of the SQRT Function

Type up the program and execute it. Some compilers will properly declare the line:

SI = SQRT(2)

to be in error. Other compilers, however, will generate an executable program. When that program is run, the result will be:

The square root of 2
1.41421   1.41421   0.00000

It can be seen that an incorrect result of zero is given for the third item.

FORTRAN PROGRAM: ENTERING DATA FROM THE KEYBOARD

Several programs in this book require input from the console keyboard. The program given in Figure 1.11 can be used to explore this aspect of a FORTRAN compiler. Type up the program and execute it. In some cases a dollar sign can be placed at the end of statements 101 and 104 so that the cursor will remain on the same line as the question.

```
C         PROGRAM TEST
C
C  --  Test console input.
C
          INTEGER N, INOUT
          REAL X
C
          INOUT = 1
10        WRITE(INOUT, 101)
          READ(INOUT, 102) N
          WRITE(INOUT, 103) N, N
          WRITE(INOUT,104)
          READ(INOUT, 105) X
          WRITE(INOUT, 106) X
          GOTO 10
101       FORMAT(' Input an integer: ')
102       FORMAT(I2)
103       FORMAT(1X,I5,' = ', A1)
104       FORMAT(' Input a real number: ')
105       FORMAT(E10.0)
106       FORMAT('   X = ', 1PE15.5)
          END
```

Figure 1.11: Test the Console Input

You will first be asked to input an integer. Respond with the value of 70. The program should repeat the value of 70 and also display the character equivalent of the number. For a byte-oriented ASCII machine this will be the letter F. The second part requests a real number; respond with the value of 1.23. The program will display the real number in an E format. (See Appendix A for a summary of format descriptors.) It will then repeatedly cycle through these two operations. For the next cycle, respond to the first question with the single digit 7. While the format for this READ statement is I2, most interactive FORTRANs will understand that the number seven is desired. However, there is at least one compiler which fills out format fields with zeros. In this case the value of 70 will be displayed. On the next pass, give the value of 07. If the console bell sounds at this step, you have found one method of giving a signal to the operator. This is because the literal equivalent of 7 corresponds to the ASCII bell character.

Pay particular attention to the format of the real number. Try values such as 12 (without a decimal point) and 1.2345E−2. Here, too, most interactive compilers will generate the desired value. Unfortunately, however, at least one compiler will give a very different result. In this case, the decimal point must be included and the integer part following the E format must be right-adjusted in the field. Three examples are given in Figure 1.12. The second and third items may be incorrectly interpreted.

Input Value	Literal Interpretation	Correct Interpretation
12.0	1.20000E+01	1.20000E+01
12	1.20000E+09	1.20000E+01
1.2345E−2	1.23450E−20	1.23450E−02

Figure 1.12: Two Interpretations of Console Input for an E10.0 Format

SUMMARY

In this chapter we have written several evaluative tools designed for testing FORTRAN compilers. We began with a program to test floating-point operations; we ran this program on two different compilers and compared the results. We then discovered an interesting quirk in the SIN function of some FORTRAN compilers, and we devised a program for testing this function. Next, we wrote a program to explore the double precision floating-point operations. We considered the consequences of mixing real and integer values for function and subroutine arguments. We saw that some compilers give incorrect results in this case. Finally, we saw how a compiler might misinterpret console input.

EXERCISES

1-1: *Write a routine to test the square root function, SQRT. Input a value from the keyboard. Take the square root, and print the result. Square this value and print the answer.*

1-2: *Write a statement function called TAN(X) to calculate the tangent of an angle X. Use the intrinsic functions SIN and COS. Test this tangent function and the intrinsic arctangent function ATAN by writing a program to take the combinations TAN(ATAN(X)) and ATAN(TAN(X)).*

1-3: *Write a program to read values from the keyboard and take EXP(ALOG(X)) and ALOG(EXP(X)).*

2

Mean and Standard Deviation

IN THIS CHAPTER we will review some statistical tools and discuss how to implement these tools in FORTRAN. We will describe the uses of the mean and the standard deviation, and present a program that calculates both of these values. We will then discuss random numbers and some methods of generating them on a computer. In particular, we will look at two FORTRAN implementations of random number generators. Finally, we will evaluate both of these random functions, using FORTRAN programs designed to test the "randomness" of the resulting number.

THE MEAN

We often use a single number, called the *average* or *mean value*, to summarize a particular group of data. The mean is calculated by adding all the items in the group and then dividing this sum by the number of items. The formula is:

$$\bar{y} = \frac{\sum_{i=1}^{N} y_i}{N} \qquad (1)$$

where y_i is the set of data containing N elements. The symbol, $\bar{y}$, (pronounced y-bar) is the resulting mean.

On the other hand, when there is a uniform *continuous distribution* of the data, the mean can be determined by integration. Consider the function $y = f(x)$ over the interval from limit a to limit b. The mean value of y is constant over this interval. Consequently, the area under the mean will be equal to the area under the curve $f(x)$:

$$\bar{y}\,(b - a) = \int_a^b f(x)dx$$

Therefore, the average value of y is:

$$\bar{y} = \frac{\int_a^b f(x)dx}{b - a}$$

Dispersion from the Mean

Sometimes all the data are close to the mean. In other cases, there is a great range of values. As an example of the latter, consider the reporting of weather. The average annual rainfall of San Francisco is said to be 19 inches and the average annual snowfall of New York City is given as 30 inches. But some years are very wet and others are very dry. The average annual temperature of Albuquerque and San Francisco is exactly the same, 57 degrees, but the climate of these two cities is very different.

As another example, consider a particular brand of breakfast cereal that contains the statement:

Net weight 16 ounces

on the box. Suppose that an inspector from the Bureau of Weights and Measures decides to check this brand of cereal in a grocery store. Several boxes are opened and the contents are weighed. Some boxes are found to contain exactly 16 ounces, but others have 15 ounces or 17 ounces. Should the boxes that contain only 15 ounces be confiscated as examples of short weight? If the contents of 100 boxes of cereal are weighed, the resulting frequency distribution might look like the curve in Figure 2.1.

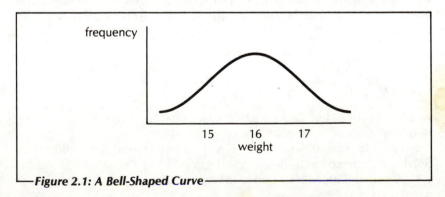

Figure 2.1: A Bell-Shaped Curve

The average weight is 16 ounces, but some of the cereal boxes weigh more than the mean value and others weigh less. Furthermore, there are very few boxes that weigh more than 17 ounces or less than 15 ounces.

The frequency distribution shown in Figure 2.1 is *bell shaped*. The curve shows a *Gaussian* or *normal* distribution. This behavior is typical of random variation about a mean value. The equation of this bell-shaped curve is related to the Gamma function and the Gaussian error function, which we will study in Chapter 11.

Now, suppose that a second type of breakfast cereal is also tested. The results this time might look like the curve in Figure 2.2. The data again show a frequency distribution that is bell shaped with a mean value of 16 ounces. But this time, there is a larger dispersion in weights. Some boxes are as heavy as 26 ounces while others are as light as 6 ounces.

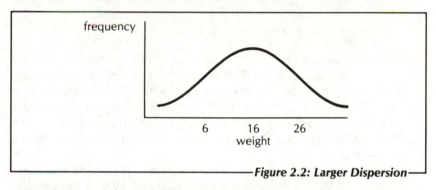

Figure 2.2: Larger Dispersion

Clearly, there is a difference in packaging between the first type of cereal and the second, even though they both have the same average weight. Something else besides the mean value is needed to describe the distribution. We need something that describes the *dispersion*. The tool that we can use, the standard deviation, is described in the next section.

THE STANDARD DEVIATION

The *standard deviation* is a measure of dispersion about the mean value. A large standard deviation means a large dispersion; a small deviation means a small dispersion. The symbol for the standard deviation is the lower-case Greek letter sigma. The standard deviation is defined by the relationship:

$$\sigma = \sqrt{\dfrac{\sum\limits_{i=1}^{N}(\overline{y} - y_i)^2}{N - 1}} \tag{2}$$

where $\overline{y}$ is the mean and y_i is the set of N data.

Equation 2 demonstrates the meaning of the standard deviation. The square of the difference between each element and the mean value is important. If the elements are closely grouped about the mean, then this difference will be small. The corresponding standard deviation will also be small. On the other hand, if data are spread far from the mean, this difference will be large. The resulting standard deviation will also be large.

The standard deviation can be used quantitatively to describe the dispersion of a set of data. For example, a range of one standard deviation on either side of the mean includes about 68% of the sample. About 95% of the values lie within two standard deviations of the mean, and almost all the values, 99.7%, lie within three standard deviations of the mean.

$$\bar{y} \pm \sigma \qquad 68\%$$

$$\bar{y} \pm 2\sigma \qquad 95\%$$

$$\bar{y} \pm 3\sigma \qquad 99.7\%$$

Suppose that we want to select a type of steel for constructing a bridge. Tensile tests are conducted to determine the strength. One steel is found to have a strength of 450 MPa (megapascals), with a standard deviation of 10 MPa. The results indicate that 99.7% of the pieces are expected to have a strength in the range 420 to 480 MPa, that is, within three sigmas. Consequently, a design could be based on a strength of 420 MPa, three sigmas below the mean value.

But suppose that a second type of steel was found to have a mean strength of 500 MPa. Is it a better steel than the first? What if the standard deviation for this second steel is 40 MPa? This larger sigma means a larger spread in values. A strength of three sigmas below the mean of this second steel is only 380 MPa. This second steel is, therefore, not as good, even though its mean value is higher.

Metals used for construction typically have small sigmas, but brittle materials, such as concrete and glass, usually have large sigmas. Suppose that tests are made on a type of glass suitable for an office door. The average breaking strength is 120 MPa. But if the sigma is only 40 MPa, then three sigmas below the mean gives a value of zero. Thus, it is important to consider the standard deviation as well as the mean value.

In this section we have learned the meaning and the importance of the standard deviation. We will now see how to calculate this value and we will design a FORTRAN program to perform the calculation.

Calculation of the Standard Deviation

While Equation 2 correctly demonstrates the meaning of the standard deviation, it is an inferior method of calculation. The subtraction of each element from the mean value will cause a loss of significance. This is especially important for points that are very close to the mean. That is, the smaller the value of sigma, the greater the problem will be. Another disadvantage of Equation 2 is that a calculation of the average is required before any subtractions can be performed. This calculation cannot be performed until all the data are available.

A better method of calculating the standard deviation can be derived by expanding the numerator of Equation 2:

$$\Sigma \, (\bar{y}^2 - 2\bar{y}y_i + y_i^2)$$

which reduces to

$$\left(\frac{\Sigma y_i}{N}\right)^2 N - 2\left(\frac{\Sigma y_i}{N}\right) \Sigma y_i + \Sigma y_i^{\,2}$$

since $\bar{y} = \Sigma y_i / N$. The resulting formula is:

$$\sigma = \sqrt{\frac{\Sigma y^2 - \Sigma y \, \Sigma y / N}{N - 1}} \tag{3}$$

With this method, two running totals are kept: the sum of the individual values and the sum of the squares of the values.

FORTRAN PROGRAM: MEAN AND STANDARD DEVIATION

The program shown in Figure 2.3 can be used to calculate the mean and standard deviation of a set of numbers. The main program performs several tasks:

1. Declare the scalar variables and the array X.

2. Print the title.

3. Call three subroutines: INPUT to ask the user for data, MEANST to calculate the mean and standard deviation, and OUTPUT to print the results.

Subroutine MEANST finds the sum of the values (SUM) and the sum of the squares of the values (SUMSQ) in a loop containing the two lines:

```
SUM = SUM + X(I)
SUMSQ = SUMSQ + X(I)*X(I)
```

Immediately following this loop, the mean (MEAN) and standard deviation (STD) are calculated.

```
MEAN = SUM / LENGTH
STD = SQRT((SUMSQ - SUM*SUM/LENGTH)/(LENGTH - 1))
```

These two lines correspond to Equations 1 and 3 above.

Subroutine MEANST demonstrates a feature we will frequently use in this book. Parameters in the heading of a subroutine, such as X, LENGTH, MEAN, and STD in subroutine MEANST, are dummy variables. They cannot be initialized by the compiler. Furthermore, a dummy array such as X has no intrinsic length. Rather, it takes the length of the corresponding array of the calling program. Therefore, the declared length in a DIMENSION or REAL statement is of no importance. Throughout this book, all one-dimensional dummy arrays will generally be given a length of unity in the dimension statement. This will be a reminder that they are dummy variables.

The program begins by asking the user for the number of values to be entered. If the answer exceeds the declared length of the array X, then the question is asked again. After the values have been entered, the mean and standard deviation are determined.

If many values are to be entered, an alternate method might be considered. For example, the computer program could count the number of values as they are entered. A special value, such as a number less than −20000, could be used to signal the end of the list.

```
C       PROGRAM MEANS
C
C -- Find the mean and standard deviation.
C -- April 30, 81
C
        INTEGER IN, OUT, LENGTH, MAX
        REAL X(80), MEAN, STD
        COMMON /INOUT/ IN, OUT
C
        IN = 1
        OUT = 1
        MAX = 80
        WRITE(OUT, 101)
10      CALL INPUT(X, LENGTH, MAX)
        IF (LENGTH .LT. 2) GOTO 99
        CALL MEANST(X, LENGTH, MEAN, STD)
        CALL OUTPUT(X, LENGTH, MEAN, STD)
        GOTO 10
99      STOP
101     FORMAT('1 Calculation of mean',
     *  ' and standard deviation')
        END
        SUBROUTINE INPUT(X, N, MAX)
C -- Get values for N and array X.
C
        INTEGER IN, OUT, I, N, MAX
        REAL X(1)
        COMMON /INOUT/ IN, OUT
C
10      WRITE(OUT, 101)
```

Figure 2.3: Calculation of the Mean and Standard Deviation

```
           READ(IN, 102) N
           IF (N .GT. MAX) GOTO 10
           IF (N .LT. 2) RETURN
           DO 20 I = 1, N
              WRITE(OUT, 103) I
              READ(IN, 104) X(I)
  20          CONTINUE
           RETURN
  101      FORMAT(' How many points? ')
  102      FORMAT(I2)
  103      FORMAT('+', I3, ':')
  104      FORMAT(F10.0)
           END
           SUBROUTINE OUTPUT(X, N, MEAN, STD)
C
C -- Print the answers.
C
           INTEGER IN, OUT, N, I
           REAL X(1), MEAN, STD
           COMMON /INOUT/ IN, OUT
C
           WRITE(OUT, 101) N, MEAN, STD
           RETURN
  101      FORMAT(' For', I3, ' points, mean =', F8.4,
      *    ', sigma =', F8.4)
           END
           SUBROUTINE MEANST(X, LENGTH, MEAN, STD)
C
C -- Find the mean and standard deviation.
C
           INTEGER LENGTH, I
           REAL X(1), MEAN, STD, SUM, SUMSQ
C
           SUM = 0.0
           SUMSQ = 0.0
           DO 10 I = 1, LENGTH
              SUM = SUM + X(I)
              SUMSQ = SUMSQ + X(I)*X(I)
  10          CONTINUE
           MEAN = SUM / LENGTH
           STD = SQRT((SUMSQ - SUM*SUM/LENGTH)/(LENGTH - 1))
           RETURN
           END
```

Figure 2.3: Calculation of the Mean and Standard Deviation (cont.)

Type up the program and run it. Input the 5 numbers 1,2,3,4, and 5, and verify that the mean is 3 and the standard deviation is 1.58. The results should look like Figure 2.4. Notice that a plus symbol is used in statement 103 of the input subroutine. This is one way to suppress an additional line after the user enters each item of data.

```
Calculation of mean and standard deviation
How many points? 5
1:  1
2:  2
3:  3
4:  4
5:  5

For  5  points, mean =  3.0000  sigma =  1.5811
```

Figure 2.4: Output from the Mean and Standard Deviation Program

This completes our introduction to the mean and the standard deviation. We will later return to these subjects; but we will now proceed to the subject of random numbers and random number generators.

RANDOM NUMBERS

A set of random numbers can sometimes be used to test a computer program if actual data are not available. For example, suppose that we have written a program to fit a straight line through a set of experimental data. We could generate a set of points along a straight line. Then, we could selectively move the points off the line by using a random number generator.

FORTRAN FUNCTION: A RANDOM NUMBER GENERATOR

A random number generator is not usually incorporated into standard FORTRAN. However, one may be provided with your FORTRAN package. Random numbers are required for some of the programs given in this book. If your FORTRAN does not include a random number generator, then you can use the function subprogram shown in Figure 2.5 for this purpose. The function contains a single dummy parameter that is not actually used in the generation of the random numbers. Its inclusion, however, simplifies the calling procedure. Furthermore, when random number generators are included in FORTRAN packages, they frequently utilize an argument to reset the seed.

This random number generator will return a sequence of numbers between zero and unity. The function utilizes the equation:

$$x = (\text{seed} + 3.141593)^5$$

to produce a sequence of random numbers. The seed is added to the value of π and the fifth power is taken. The fractional part is used for both the next random number and the seed for the following number. A DATA statement near the beginning of function RAN is used to initialize the

```
        FUNCTION RAN(DUMMY)
C
C  --  Generate random numbers from 0 to 1.
C
        REAL SEED, X
        DATA SEED/4.0/
C
        X = (SEED + 3.141593)**5
        SEED = X - AINT(X)
        RAN = SEED
        RETURN
        END
```

Figure 2.5: A FORTRAN Subprogram for Generating Random Numbers

random number seed during the compiling step. Since the DATA statement is not executable, it is ignored during program execution.

Now that we have learned how our random number generator works, let us look at a program that tests the quality of the random number generator.

FORTRAN PROGRAM:
EVALUATION OF A RANDOM NUMBER GENERATOR

Whether you use the random number generator given in Figure 2.5, or the one supplied with your FORTRAN, you should perform a simple test to see how reasonable the results are. The mean value of a set of uniformly distributed random numbers ranging from zero to unity should, of course, be one-half. Furthermore, the standard deviation should be the reciprocal of the square root of 12, a value of 0.2887.

The program shown in Figure 2.7 can be used to test a random number generator. Your random number generator may need to be initialized by calling it with a negative argument. A separate instruction has been included for this purpose. It is not needed, however, if you use the random number generator shown in Figure 2.5. The random number routine is called 48 times from the inner DO loop. The random numbers are stored in the array X. Subroutine MEANST is then called to calculate the mean and standard deviation. Both the subroutine call and a print statement are contained in the outer DO loop which repeats the process 20 times.

Running the Program

Type up the program shown in Figure 2.7. The random number generator given in Figure 2.5 should also be included if necessary. Now test your random number generator. Random number generators are sometimes called pseudo-random number generators to emphasize the fact that they are not truly random. You should not expect to obtain a

mean of one-half and a standard deviation of 0.2887 for each grouping of
48 values. The first part of the output from the program might look like
Figure 2.6.

```
Test of random number generator
      mean          std dev
     (0.5)         (0.2887)
    ====================
    .5458          .3046
    .4543          .2785
    .5075          .2740
    .5443          .3152
    .4952          .2803
    .4814          .2963
    .4569          .3115
    .5103          .2587
    .5447          .2745
    .5079          .2690
    .4836          .2754
    .5193          .2897
    .4504          .2748
    .4377          .2851
    .5429          .2976
    .5497          .2853
    .5504          .3023
    .4982          .3241
    .4721          .2835
    .5332          .2535
```

Figure 2.6: Output: Testing a Random Number Generator

Notice that the mean ranges from 0.44 to 0.55, and the standard deviation
has values from 0.25 to 0.32. Other random number generators may be
better or worse than this.

```
C        PROGRAM RANDT
C
C -- Program to test the random number generator.
C -- MAY 22, 81
C
        INTEGER OUT, LENGTH, I, J
        REAL X(100), MEAN, STD, SEED
C
        OUT = 1
        MAX = 100
        LENGTH = 48
        WRITE(OUT, 101)
C -- Reset the random number generator.
        DUMMY = RAN(-1.0)
```

Figure 2.7: Program to Test the Random Number Generator

```
        DO 20 I = 1, 20
          DO 10 J = 1, LENGTH
            X(J) = RAN(1.0)
10        CONTINUE
          CALL MEANST(X, LENGTH, MEAN, STD)
          WRITE(OUT, 102) MEAN, STD
20      CONTINUE
        WRITE(OUT, 103)
        STOP
101     FORMAT('1 Test of random number generator'/
     *   '       mean        std dev'/
     *   '       (0.5)        (0.2887)'/
     *   '       ==================')
102     FORMAT(1X, 2F10.4)
103     FORMAT(/)
        END
        SUBROUTINE MEANST(X, LENGTH, MEAN, STD)
C
C -- Find the mean and standard deviation.
C
        INTEGER LENGTH, I
        REAL X(1), MEAN, STD, SUM, SUMSQ, XI
C
        SUM = 0.0
        SUMSQ = 0.0
        DO 10 I = 1, LENGTH
           XI = X(I)
           SUM = SUM + XI
           SUMSQ = SUMSQ + XI*XI
10      CONTINUE
        MEAN = SUM / LENGTH
        STD = SQRT((SUMSQ - SUM*SUM/LENGTH)/(LENGTH - 1))
        RETURN
        END
```

Figure 2.7: Program to Test the Random Number Generator (cont.)

Using Pi to Produce Random Numbers

A short sequence of random numbers can be obtained from the first 15 digits of pi:

3.14159265358979

The following mnemonic makes it easy to remember the sequence. The number of letters in each word is equal to the corresponding digit of pi:

YES, I NEED A DRINK, ALCOHOLIC, OF COURSE, AFTER

3. 1 4   1 5   9   2   6   5

THE HEAVY SESSIONS INVOLVING QUANTUM MECHANICS

3   5   8   9   7   9

This sequence has a mean value of 5.1 and a standard deviation of 2.8.

Our second type of random number generator, described in the next section, is designed to simulate experimental data.

Gaussian Random Numbers

Sampling errors, that is, errors made during measurement, can be of any size. However, small errors are more likely than large errors. For example, suppose that a table with an actual length of six feet is measured with a ruler. An error of one foot is less likely than an error of one inch. An error of ten feet is even less likely. Thus, measured values that are further from the correct value are less likely than values that are closer. Consequently, if we want to simulate experimental data with a random number generator, then the numbers should not be uniformly spaced. But in fact, the usual random number generator produces a uniform set of numbers over the interval 0 to 1.

What is needed for the simulation of experimental data is a set of random numbers that are grouped about the mean. Thus, the frequency distribution should be bell shaped; it should have a normal or Gaussian distribution. Fortunately, it is fairly easy to produce a Gaussian distribution of random numbers from an ordinary random number generator.

Consider a sequence of 12 random numbers. It is highly unlikely that all 12 will have a value of zero. Similarly, it is very unlikely that they will all be unity. Suppose that we generate a new number from the sum of 12 uniformly distributed random numbers. Since the average value of the original numbers is one-half, the sum of 12 random numbers is likely to be about one half of 12, or 6. Thus, Gaussian random numbers can be generated from an ordinary random number generator. We will now look at a FORTRAN implementation of such a function.

FORTRAN PROGRAM:
GENERATING AND TESTING GAUSSIAN RANDOM NUMBERS

Gaussian random numbers can be obtained by summing 12 uniformly distributed random numbers and subtracting the value of 6. A set of such numbers should have a mean value of zero and a standard deviation of unity. Other values for the mean and standard deviation can readily be chosen. The formula for calculating each random number, RANDG, is:

$$RANDG = (SUM - 6) * SIGMA + MEAN$$

In this expression, SUM is the sum of 12 uniformly distributed random numbers, SIGMA is the desired standard deviation, and MEAN is the desired mean. If each Gaussian random number is formed from more than 12 numbers, then there is an additional complication. In this case, the formula becomes:

$$RANDG = (SUM - NUM/2) * SIGMA * SQRT(12/NUM) + MEAN$$

where NUM is the number of uniformly distributed random numbers used to obtain each Gaussian random number.

A Gaussian random number generator can be combined with the regular random number generator to produce a series of Gaussian-distributed random numbers. There are two parameters that must be supplied by the calling program: the desired mean and the standard deviation of the resulting numbers. An alternative approach would be to set these values in the subroutine.

A FORTRAN program for generating and testing a sequence of Gaussian random numbers is given in Figure 2.8. The main program sets the desired mean (MEAN) to 10 and the desired standard deviation (SIGMA) to 0.5. The outer loop that controls the subroutine calls is the same as the one given in Figure 2.7. The inner subroutine, however, now calls function RANDG 50 times to produce an array of Gaussian random numbers.

```
C       PROGRAM RANDGT
C
C -- Program to test Gaussian random number generator.
C -- Function RAN and subroutine MEANST are required.
C -- May 22, 81
C
        INTEGER OUT, LENGTH, I, J
        REAL X(100), MEAN, STD, DMEAN, DSTD
C
        OUT = 1
        MAX = 100
        LENGTH = 50
        DMEAN = 10.0
        DSTD = 0.5
C -- Reset the random number generator.
        DUMMY = RAN(-1.0)
        WRITE(OUT, 101)
        DO 20 I = 1, 20
          DO 10 J = 1, LENGTH
            X(J) = RANDG(DMEAN, DSTD)
10        CONTINUE
          CALL MEANST(X, LENGTH, MEAN, STD)
          WRITE(OUT, 102) MEAN, STD
20      CONTINUE
        WRITE(OUT, 103)
99      STOP
101     FORMAT('1 Generation of Gaussian random numbers'/,
     *   '        mean       std dev'/
     *   '        (10)        (0.5)'/
     *   '        ==================')
```

Figure 2.8: Generating and Testing Gaussian-Distributed Random Numbers

```
102      FORMAT(1X, 2F10.4)
103      FORMAT(/)
         END
         FUNCTION RANDG(MEAN, SIGMA)
C
C --  Generate Gaussian random numbers with
C --  requested mean and standard deviation.
C --  Function RAN is required.
C
         INTEGER I
         REAL SUM, MEAN, SIGMA
C
         SUM = 0.0
         DO 10 I = 1, 12
            SUM = SUM + RAN(1.0)
10       CONTINUE
         RANDG = (SUM - 6) * SIGMA + MEAN
         RETURN
         END
```

Figure 2.8: Generating and Testing Gaussian-Distributed Random Numbers (cont.)

The area under the normal distribution curve is related to the standard deviation. It can be obtained from the Gaussian error function that is developed in Chapter 11.

SUMMARY

We began this chapter with a discussion of two important statistical tools, the mean and the standard deviation. Our discussion led us to a FORTRAN program for calculating these values. Then we discovered how to write two different FORTRAN random number generators, which will prove to be useful tools if our system does not supply such a function. Finally, we discussed the importance of evaluating the reasonableness of the generated random numbers and we saw two programs that will carry out that evaluation.

In the next chapter we will continue to expand our understanding of FORTRAN; we will see how to express vectors and matrices as arrays.

EXERCISES

2-1: *Alter the program given in Figure 2.3 so that the number of items to be averaged is determined by the program. Use a value less than −20000 as a signal for the end of data. The program should repeatedly loop, receiving items and incrementing the number of items, N, until a number less than −20000 is entered.*

2-2: *Alter the program given in Figure 2.3 so that the items to be averaged are obtained from a DATA statement in the program rather than from the keyboard. Use the value −20000 as a signal for the end of data. The program should repeatedly loop, receiving items and incrementing the number of items, N, until a number smaller than −20000 is encountered.*

2-3: *Verify that the first 15 digits of pi have an average value of 5.1 and a standard deviation of 2.8.*

2-4: *Alter the program given in Figure 2.8 so that 24 uniformly distributed random numbers are used to generate each Gaussian random number.*

3

Vector and Matrix Operations

IN MOST OF THE CHAPTERS of this book, we develop programs that use vectors and matrices. Consequently, in this chapter, we will review the concepts of vectors and matrices and consider some of their more common mathematical operations. We will demonstrate several FORTRAN implementations. Vectors and matrices are important concepts because they greatly simplify the programming of mathematical operations on data sets.

SCALARS AND ARRAYS

We will begin our discussion by establishing the difference between a *scalar* variable and an *array*. An ordinary, simple variable is called a scalar. It is referenced by a unique, symbolic name, and it is associated with a single value. For example, the FORTRAN expression:

    YEAR = 1971

assigns a value to the scalar variable YEAR. In FORTRAN, scalars can be declared as real, integer, complex, double-precision (real), or logical (boolean) variables.

Sometimes it is necessary to refer collectively to a set of scalar values. FORTRAN uses the *array* for this purpose. In an array, all the components have the same type. For example, they might be real numbers or integers.

The components of an array are collectively referenced by a symbolic name. Each position of the array is uniquely designated by an index or subscript that follows the name. The corresponding value at each position in the array can be individually accessed through this index. The value can be changed without affecting the other values of the array.

In the following sections we will see how *vectors* and *matrices* are represented as arrays in FORTRAN programs. We will begin by describing vectors.

VECTORS

A vector is a one-dimensional array. (The number of dimensions refers to the number of subscripts, not the number of elements.) Each element of a vector is referenced through a single index that runs from 1 up to the maximum number of elements in the vector.

In ordinary usage, the elements of a vector are separated by spaces. For example, consider the vector **v** that contains the values:

251943

The FORTRAN statement:

REAL V(6)

can be used to define the symbol V as a vector of real numbers. The maximum number of elements (the length) is declared to be six. Values can be assigned to this vector by FORTRAN statements such as:

V(1) = 2
V(2) = 5
V(3) = 1
V(4) = 9
V(5) = 4
V(6) = 3

The first element of this vector is located at position V(1), the second element is at V(2) and the last element is at V(6). The first element of this vector has a value of 2, the second element has a value of 5 and the last element has a value of 3.

Since vectors have only a single dimension, it should not matter whether they are written horizontally or vertically. But sometimes we will need to distinguish between *row vectors*, which are written horizontally, and *column vectors*, which are written vertically. For example, the set:

$$\begin{bmatrix} 2 \\ 5 \\ 1 \\ 9 \\ 4 \\ 3 \end{bmatrix}$$

is a column vector.

In the next section we will study the operations of vector arithmetic and their implementation in FORTRAN.

Vector Arithmetic

The major arithmetic operations defined for vectors are: magnitude, scalar multiplication, vector addition, dot product, and cross product. Let us now look at each of these operations.

Magnitude

The *magnitude* of a vector is a scalar value. It is obtained by summing the squares of the elements, and then taking the square root of the resulting sum. For example, the magnitude of the vector **y**:

$$\mathbf{y} = \begin{bmatrix} 2 & 2 & 1 \end{bmatrix}$$

is equal to the square root of $4 + 4 + 1$. The resulting value is 3. The operation can be performed with the function statement:

$$MAG(Y) = SQRT(Y(1)*Y(1) + Y(2)*Y(2) + Y(3)*Y(3))$$

Scalar Multiplication of Vectors

If a vector **y** is multiplied by a scalar value *s*, then each element of **y** is multiplied by *s*. For example, 2 times the vector **y** is:

$$2\mathbf{y} = \begin{bmatrix} 4 & 4 & 2 \end{bmatrix}$$

The following FORTRAN statements generate a vector Y2 in which each element is twice as large as the corresponding element of vector Y:

```
      DO  10  I = 1, 6
10    Y2 (I)  = 2.0 * Y(I)
```

Vector Addition

Two vectors can be added together if each has the same number of elements, or the same length. The result is a new vector in which each element is the sum of the two corresponding elements of the original vectors.

Thus, if:

$$\mathbf{a} = [1 \quad 2 \quad 3]$$

and

$$\mathbf{b} = [3 \quad 4 \quad 5]$$

then

$$\mathbf{a} + \mathbf{b} = [4 \quad 6 \quad 8]$$

The corresponding FORTRAN expression is:

```
      DO   10   I = 1, 3
10    C(I) = A(I) + B(I)
```

Dot Product (or Scalar Product)

The *dot product* or *scalar product* of two equal-length vectors produces a scalar result. Each element of one vector is multiplied by the corresponding element of the other. The resulting products are then added together. The mathematical symbol for the dot operator is simply a dot placed between the operands. Thus:

$$\mathbf{a} \cdot \mathbf{b} = (1)(3) + (2)(4) + (3)(5) = 26$$

The dot product is equal to the product of the magnitude of the original vectors and the cosine of the angle between them:

$$\mathbf{a} \cdot \mathbf{b} = |\mathbf{a}| \, |\mathbf{b}| \cos \theta$$

This formula can be used to find the angle between two vectors. For example, the angle between the vectors A and B is 10.7°:

$$\cos \theta = 26/ \, (3.742 \cdot 7.071) = 0.9826 \quad \text{and}$$
$$\text{Arccos } 0.9826 = 10.7°$$

Cross Product (or Vector Product)

The *cross product*, or *vector product*, of two vectors produces a third vector that is mutually perpendicular to the two original vectors. The mathematical symbol for this operation is an $\times$. The magnitude of the resulting vector is equal to the product of the magnitudes of the original vectors and the sine of the angle between them:

$$|\mathbf{a} \times \mathbf{b}| = |\mathbf{a}| \, |\mathbf{b}| \sin \theta$$

The cross product, $\mathbf{c} = \mathbf{a} \times \mathbf{b}$, can be calculated from the FORTRAN expressions:

$$C(1) = A(2)*B(3) - B(2)*A(3)$$
$$C(2) = -A(1)*B(3) + B(1)*A(3)$$
$$C(3) = A(1)*B(2) - B(1)*A(2)$$

The cross product of the vectors **a** and **b** is the vector $[-2 \ 4 \ -2]$; its magnitude is 4.9.

We have now seen FORTRAN implementations for the main arithmetic operations on vectors. Later in this chapter we will see the matrix equivalents of these operations.

MATRICES

A two-dimensional array is called a *matrix*. The elements of this set are arranged into a rectangle or square. The elements of the matrix can be considered as a set of horizontal lines called row vectors or as a set of vertical lines called column vectors. A matrix can thus be described as a one-dimensional set of vectors.

Each element of a matrix is uniquely defined by a pair of indices: the *row index* and the *column index*. For example, consider the matrix:

$$\begin{bmatrix} x_{11} & x_{12} & x_{13} & \cdots & x_{1m} \\ x_{21} & x_{22} & x_{23} & \cdots & x_{2m} \\ x_{31} & x_{32} & x_{33} & \cdots & x_{3m} \\ \cdots & \cdots & \cdots & \cdots & \cdots \\ x_{n1} & x_{n2} & x_{n3} & \cdots & x_{nm} \end{bmatrix}$$

which contains n rows and m columns. The row index is always given first. It is then followed by the column index.

A matrix is referenced by its name, which can be a single alphabetic character, or a string of characters. The indices are given as subscripts except in computer programs, where subscripting is not possible. In COBOL, FORTRAN, and BASIC programs, the array subscripts are enclosed in parentheses. Square brackets are used for this purpose in Pascal and APL. Thus, the appearance of the expression X(2,3) in a FORTRAN program is a reference to the element located in the second row and the third column of the two-dimensional array named X.

A matrix that has the same number of rows as columns is called a *square matrix*. The *principal*, or *main diagonal* of a square matrix contains the elements $x_{11}, x_{22}, x_{33}, \ldots, x_{nn}$. The principal diagonal is sometimes called simply the *diagonal*. A square matrix that contains the value of unity at each position of the main diagonal, and is zero everywhere else, is known as a *unit matrix,* or an *identity matrix.* For example:

$$\begin{bmatrix} 1 & 0 & 0 & 0 \\ 0 & 1 & 0 & 0 \\ 0 & 0 & 1 & 0 \\ 0 & 0 & 0 & 1 \end{bmatrix}$$

is a 4-by-4 unit matrix.

Now that we have defined the essential vocabulary of matrices, we can go on to study the major arithmetic operations for matrices, and the FORTRAN implementations of these operations.

Matrix Arithmetic

We will begin by defining the transpose operation; then we will describe scalar multiplication, and matrix addition, subtraction, and multiplication.

The Transpose Operation

A matrix is *transposed* by interchanging the rows and the columns. Each original element x_{ij} becomes the new element x_{ji} of the transposed matrix. The *transpose* of matrix X is designated as X^T. Thus, if

$$X = \begin{bmatrix} 1 & 2 & 3 \\ 4 & 5 & 6 \\ 7 & 8 & 9 \end{bmatrix}$$

then,

$$X^T = \begin{bmatrix} 1 & 4 & 7 \\ 2 & 5 & 8 \\ 3 & 6 & 9 \end{bmatrix}$$

Notice that the transpose of a square matrix can be obtained by rotation of the matrix about the principal diagonal.

Two matrices are *equal* if every element of one is equal to the corresponding element of the other. Thus if $X = Y$, then for each element:

$$x_{ij} = y_{ij}$$

A square matrix is symmetric if it is equal to its transpose. In this case, each element x_{ij} equals the corresponding element x_{ji}.

Scalar Multiplication of Matrices

If a matrix X is multiplied by a scalar value s, then each element of the matrix is multiplied by the value *s*. The following FORTRAN statements will produce a matrix Y from the product of matrix X and the scalar S.

```
DO    10   I = 1, N
    DO   10   J = 1, M
10   Y(I,J) = X(I,J) * S
```

Matrix Addition and Subtraction

One matrix may be added to or subtracted from another matrix if both have the same number of columns and the same number of rows. The addition of the matrix X to the matrix Y to produce matrix Z is written as:

$$Z = X + Y$$

Thus, if

$$X = \begin{bmatrix} x_{11} & x_{12} & x_{13} \\ x_{21} & x_{22} & x_{23} \end{bmatrix}$$

and

$$Y = \begin{bmatrix} y_{11} & y_{12} & y_{13} \\ y_{21} & y_{22} & y_{23} \end{bmatrix}$$

then

$$Z = \begin{bmatrix} x_{11} + y_{11} & x_{12} + y_{12} & x_{13} + y_{13} \\ x_{21} + y_{21} & x_{22} + y_{22} & x_{23} + y_{23} \end{bmatrix}$$

The corresponding FORTRAN statements are:

```
DO    10   I = 1, N
    DO   10   J = 1, M
10   Z(I,J) = X(I,J) + Y(I,J)
```

Each element of Z is formed from the sum of the corresponding elements of X and Y. In a similar way, subtraction of one matrix from another:

$$Z = X - Y$$

is performed by subtracting each element of the second matrix from the corresponding element of the first.

Matrix Multiplication

One matrix may be multiplied by another if the number of columns of the first matrix equals the number of rows of the second. The two matrices are said to be *conformable* in this case. Thus, if X is a matrix that contains *m* rows and n columns, and Y is a matrix with n rows and p columns, then the product:

$$Z = XY$$

produces a matrix Z with m rows and p columns. That is, the resulting matrix has the same number of rows as matrix X and the same number of columns as matrix Y.

In matrix multiplication, each element of Z is formed from a sum of products. The elements from one row of matrix X are each multiplied by the corresponding elements from a column of matrix Y, then summed up. For matrix X (which might be the transpose of the previous X) and matrix Y:

$$X = \begin{bmatrix} x_{11} & x_{12} \\ x_{21} & x_{22} \\ x_{31} & x_{32} \end{bmatrix}$$

and

$$Y = \begin{bmatrix} y_{11} & y_{12} & y_{13} \\ y_{21} & y_{22} & y_{23} \end{bmatrix}$$

the operation $Z = XY$ is formed as

$$Z = \begin{bmatrix} x_{11}y_{11} + x_{12}y_{21} & x_{11}y_{12} + x_{12}y_{22} & x_{11}y_{13} + x_{12}y_{23} \\ x_{21}y_{11} + x_{22}y_{21} & x_{21}y_{12} + x_{22}y_{22} & x_{21}y_{13} + x_{22}y_{23} \\ x_{31}y_{11} + x_{32}y_{21} & x_{31}y_{12} + x_{32}y_{22} & x_{31}y_{13} + x_{32}y_{23} \end{bmatrix}$$

Each jk element of Z is formed from row j of matrix X and column k of matrix Y according to the scheme

$$z_{jk} = x_{j1}y_{1k} + x_{j2}y_{2k} + \ldots + x_{jn}y_{nk}$$

Thus, if

$$X = \begin{bmatrix} 1 & 4 \\ 2 & 5 \\ 3 & 6 \end{bmatrix}$$

and

$$Y = \begin{bmatrix} 7 & 8 & 9 \\ 10 & 11 & 12 \end{bmatrix}$$

then the product:

$$Z = XY = \begin{bmatrix} 47 & 52 & 57 \\ 64 & 71 & 78 \\ 81 & 90 & 99 \end{bmatrix}$$

is a square matrix with dimensions of 3 by 3. The first element, z_{11}, for example, is calculated as:

$$(1)(7) + (4)(10) = 47$$

Matrix multiplication is not commutative; that is, the product YX will not, in general, be equal to the product XY. Reversing the order of the previous example produces a 2-by-2 matrix, rather than a 3-by-3 matrix:

$$YX = \begin{bmatrix} 50 & 122 \\ 68 & 167 \end{bmatrix}$$

Notice that the dot product of two vectors follows the rules for matrix multiplication if the first vector is a column vector (that is, has one column), and the second is a row vector (that is, has one row).

We have seen how FORTRAN handles vector and matrix arithmetic by means of one- and two-dimensional arrays. We are now ready to write a program using several of the FORTRAN statements we have learned.

FORTRAN PROGRAM: MATRIX MULTIPLICATION

In Chapter 4, we will need a routine for matrix multiplication. Specifically, we will need to multiply the transpose of a matrix X by the original matrix X. We will also need to multiply the vector **y** by the matrix X. Therefore, we will now write a routine to perform this task.

The program shown in Figure 3.1 contains subroutine SQUARE, which will perform both of the needed operations. The main program will generate the X matrix and the **y** vector. Subroutine SQUARE will calculate the needed matrix A and vector **g** according to the equations:

$$X^TX = A$$

and

$$yX = \mathbf{g}$$

In this example, the matrix X contains 5 rows and 3 columns. Vector **y** has a length of 5.

Since the resulting matrix A is symmetric, the calculations can be simplified. For example, there will be terms in the program like:

$$A(1,3) = A(3,1)$$

Matrix A will have 3 rows (the same as the transpose of matrix X) and 3 columns (the same as matrix X). Vector **g** will have a length of 3. Actually, both **y** and **g** must be considered as row vectors, since they have a dimension of 1 row and 5 columns. Alternately, if we want to consider the vectors **y** and **g** as column vectors, then we should write the multiplication equation as:

$$y^T X = g^T$$

where the transposed column vectors become row vectors.

Type up the program shown in Figure 3.1 and execute it. The results should look like Figure 3.2. Notice that subroutine SQUARE incorporates four scalar parameters. Two of these (NROW and NCOL) designate the current matrix size for each subroutine call. The other two parameters (N and M) designate the dimensioned size of the matrices in the calling program.

Since the arrays X, Y, A, and G are dummy variables in subroutine SQUARE, they can alternately be dimensioned as:

    REAL X(N,1), Y(1), A(M,1), G(1)

```
C        PROGRAM MMULT
C
C  -- Matrix multiplication
C  -- April 27, 81
C
         INTEGER IN, OUT, NROW, NCOL, MAXR, MAXC
         REAL X(9,3), Y(9), A(3,3), G(3)
         COMMON /INOUT/ IN, OUT
C
         IN = 1
         OUT = 1
         MAXR = 9
         MAXC = 3
         CALL INPUT(X, Y, NROW, NCOL)
         CALL SQUARE(X, Y, A, G, NROW, NCOL, MAXR, MAXC)
         CALL OUTPUT(X, Y, A, G, NROW, NCOL)
         STOP
         END
         SUBROUTINE INPUT(X, Y, NROW, NCOL)
C  -- Get values for NROW, NCOL, and arrays X and Y.
C
         INTEGER I, J, NROW, NCOL
         REAL X(9,3), Y(9)
C
         NROW = 5
         NCOL = 3
         DO 20 I = 1, NROW
           X(I,1) = 1.0
           DO 10 J = 2, NCOL
             X(I,J) = I * X(I, J-1)
10         CONTINUE
           Y(I) = 2 * I
20       CONTINUE
         RETURN
         END
```

Figure 3.1: Matrix Multiplication: A = X^T X, G = Y X.

```
         SUBROUTINE OUTPUT(X, Y, A, G, NROW, NCOL)
C
C -- Print out the answers.
C
         INTEGER IN, OUT, NROW, NCOL, I, J
         REAL X(9,3), Y(9), A(3,3), G(3)
         COMMON /INOUT/ IN, OUT
C
         WRITE(OUT, 101)
         DO 10 I = 1, NROW
            WRITE(OUT, 102) (X(I,J), J = 1, NCOL), Y(I)
10       CONTINUE
         WRITE(OUT, 103)
         DO 20 I = 1, NCOL
            WRITE(OUT, 102) (A(I,J), J = 1, NCOL), G(I)
20       CONTINUE
         WRITE(OUT, 104)
         RETURN
101      FORMAT('           X                    Y')
102      FORMAT(3F8.0, ' : ', F7.0)
103      FORMAT('           A                    G')
104      FORMAT(/)
         END
         SUBROUTINE SQUARE(X, Y, A, G, NROW, NCOL, N, M)
C
C -- Matrix multiplication routine
C -- A = transpose X times X.
C -- G = Y times X.
C
         INTEGER NROW, NCOL, I, K, L, N, M
         REAL X(N,M), Y(N), A(M,M), G(M)
C
         DO 40 K = 1, NCOL
            DO 20 L = 1, K
               A(K,L) = 0.0
               DO 10 I = 1, NROW
                  A(K,L) = A(K,L) + X(I,L) * X(I,K)
                  IF (K .NE. L) A(L,K) = A(K,L)
10             CONTINUE
20          CONTINUE
            G(K) = 0.0
            DO 30 I = 1, NROW
               G(K) = G(K) + Y(I) * X(I,K)
30          CONTINUE
40       CONTINUE
         RETURN
         END
```

Figure 3.1: Matrix Multiplication: $A = X^T X$, $G = Y X$. (cont.)

```
                 X                                  Y
        1.        1.        1.    :        2.
        1.        2.        4.    :        4.
        1.        3.        9.    :        6.
        1.        4.       16.    :        8.
        1.        5.       25.    :       10.
              A                          G
        5.       15.       55.    :       30.
       15.       55.      225.    :      110.
       55.      225.      979.    :      450.
```

— Figure 3.2: Output from the Matrix Multiplication Program —

DETERMINANTS

The *determinant* of a square matrix X is designated as $|X|$. The result is a scalar value. For a 2-by-2 matrix, the upper-left member is multiplied by the lower-right, and then the product of the lower-left member and the upper-right member is subtracted:

$$|X| = x_{11}x_{22} - x_{12}x_{21}$$

For example, the determinant of the matrix:

$$\begin{bmatrix} 1 & 2 \\ 3 & 4 \end{bmatrix}$$

is -2.

The determinant of matrices larger than 2 by 2 can be found by multiplying each element of the first row by the determinant of the remaining matrix, after removing the column that is common to the element in the first row. A recursive definition is:

$$|X| = x_{11}s_{11} - x_{12}s_{12} + x_{13}s_{13} - \ldots (-1)^{n+1}x_{1n}s_{1n}$$

where x_{11}, x_{12}, etc., are the elements of the first row of matrix X, and S_{1n} is the determinant of the matrix that has row 1 and column N removed. The determinant of the matrix:

$$\begin{bmatrix} 1 & 2 & 3 \\ 4 & 5 & 6 \\ 7 & 8 & 0 \end{bmatrix}$$

is equal to:

$$1 \begin{vmatrix} 5 & 6 \\ 8 & 0 \end{vmatrix} - 2 \begin{vmatrix} 4 & 6 \\ 7 & 0 \end{vmatrix} + 3 \begin{vmatrix} 4 & 5 \\ 7 & 8 \end{vmatrix}$$

The next step is to evaluate the minor 2-by-2 matrices:

$$\begin{vmatrix} 5 & 6 \\ 8 & 0 \end{vmatrix} \quad \begin{vmatrix} 4 & 6 \\ 7 & 0 \end{vmatrix} \quad \begin{vmatrix} 4 & 5 \\ 7 & 8 \end{vmatrix}$$

according to the procedure:

$$1(5 \cdot 0 - 8 \cdot 6) - 2(4 \cdot 0 - 7 \cdot 6) + 3(4 \cdot 8 - 7 \cdot 5)$$

The resulting value of the determinant in this case is 27. If each element of a row or column is zero, then the determinant is zero. Also, if two rows or columns are identical, then the determinant is zero.

We have described the method for calculating the determinant of a matrix. Since we will be using determinants in later chapters, let us now consider a FORTRAN program that calculates determinants.

FORTRAN PROGRAM: DETERMINANTS

A program that can be used to find the determinant of a 3-by-3 matrix is given in Figure 3.3. Subroutine INPUT asks the user to enter the matrix elements. Function DETERM then calculates the value of the determinant. Type up the program and run it. The elements of the matrix are entered row by row. That is, the order is:

A(1,1), A(1,2), A(1,3),A(2,1),A(2,2),...

After the nine elements have been entered, the program displays the value of the determinant. Then the question "More?" appears. Respond with "Y" if you want to calculate the determinant of another matrix. Give an answer of "N" to terminate the program.

```
C       PROGRAM DETER
C
C -- FORTRAN program to solve three
C -- simultaneous equations by Cramer's rule.
C -- Apr 30, 81
C
        INTEGER LENGTH, YESNO, IN, OUT
        REAL A(3,3), DET
        COMMON /INOUT/ IN, OUT
C
        IN = 1
        OUT = 1
10      WRITE(OUT, 101)
        CALL INPUT(A, LENGTH)
        CALL OUTPUT(A, LENGTH)
        DET = DETERM(A)
        WRITE(OUT, 104) DET
        WRITE(OUT, 102)
        READ(IN, 103) YESNO
```

Figure 3.3: The Determinant of a 3-by-3 Matrix

```
           IF (YESNO .EQ. 'Y' .OR. YESNO .EQ. 'y') GOTO 10
           STOP
101        FORMAT('1 Determinant of a 3-by-3 matrix')
102        FORMAT(' More? ')
103        FORMAT(A1)
104        FORMAT(' The determinant is', 1PE12.4)
           END
           SUBROUTINE INPUT(A, N)
C -- Get values for N and array A.
C
           INTEGER N, IN, OUT, I, J
           REAL A(3,3)
           COMMON /INOUT/ IN, OUT
C
           N = 3
           DO 20 I = 1, N
             WRITE(OUT, 101) I
             DO 10 J = 1, N
               WRITE(OUT, 102) J
               READ(IN, 103) A(I,J)
10           CONTINUE
20         CONTINUE
           RETURN
101        FORMAT(' Equation ', I3/)
102        FORMAT('+', I4, ': ')
103        FORMAT(F10.0)
           END
           FUNCTION DETERM(X)
C
C -- Find the determinant of the 3-by-3 matrix X.
C
           REAL X(3,3), SUM
C
           SUM = X(1,1) * (X(2,2)*X(3,3) - X(3,2)*X(2,3))
           SUM = SUM - X(1,2) * (X(2,1)*X(3,3) - X(3,1)*X(2,3))
           SUM = SUM + X(1,3) * (X(2,1)*X(3,2) - X(3,1)*X(2,2))
           DETERM = SUM
           RETURN
           END
           SUBROUTINE OUTPUT(A, N)
C
C -- Print out the answers.
C
           INTEGER N, IN, OUT, I, J
           REAL A(N,N)
           COMMON /INOUT/ IN, OUT
C
           WRITE(OUT, 101) ((A(I,J), J = 1, N), I = 1, N)
           RETURN
101        FORMAT(1P3E12.4)
           END
```

Figure 3.3: The Determinant of a 3-by-3 Matrix (cont.)

INVERSE MATRICES AND MATRIX DIVISION

The *inverse* of a nonsingular matrix X is written as

$$X^{-1}$$

The inverse of a singular matrix is undefined. The product of a matrix and its inverse is the identity matrix:

$$XX^{-1} = I$$

Matrix inversion is similar to other inverse operations since the inverse of an inverse produces the original matrix:

$$(X^{-1})^{-1} = X$$

In computer implementations, however, the two will not always agree, because of roundoff error.

Matrix division is performed by a combination of inversion and multiplication. Thus, if we need to divide matrix X by matrix Y, we first perform a matrix inversion on Y. Then matrix X is multiplied by the inverse of matrix Y. The operation is written as:

$$XY^{-1}$$

Matrix inversion can be used to find the solution to a set of simultaneous linear equations. Thus, if we have a coefficient matrix A and a constant vector $\mathbf{y}$, we want a solution vector $\mathbf{b}$ such that

$$A\mathbf{b} = \mathbf{y}$$

Then the solution can be obtained from a product of the inverse of matrix A and the constant vector $\mathbf{y}$ (in that order).

$$A^{-1}\mathbf{y} = \mathbf{b}$$

Since the simultaneous solution of linear equations is the subject of the next chapter, we shall not develop a matrix inversion routine at this point.

SUMMARY

In this chapter we saw how vectors and matrices are represented in FORTRAN. We also studied the methods of performing matrix and vector arithmetic operations in FORTRAN. We developed two significant programs—one for performing matrix multiplication, and the other for calculating determinants. We will be using these routines in the chapters that follow.

EXERCISES

3-1: *Write a program that will input two (three-component) vectors from the keyboard and print the corresponding dot product. Show that the dot product of the vectors:*

      a = [1 1 1]   *and*   **b** = [1 −1 −1]

is −1.

3-2: *Write a program that will input two vectors from the keyboard and print the corresponding cross product. Show that the cross product of the vectors:*

      a = [1 1 1]   *and*   **b** = [1 −1 0]

is the vector [1 1 −2].

3-3: *Write a program that will input two vectors from the keyboard and print the corresponding angle between the two vectors. Show that the angle between the vectors:*

      a = [1 1 1]   *and*   **b** = [1 −1 −1]

is 109.5 degrees. (This is the O-Si-O bond angle in silicate minerals.)

3-4: *A unit vector has a magnitude of unity. Dividing a vector by its magnitude produces a unit vector. Write a program to convert a three-component vector, input from the keyboard, into a unit vector. Run the program to show that the vector:*

      a = [18 −18 9]

is associated with the unit vector [2/3 −2/3 1/3].

4

Simultaneous Solution of Linear Equations

IN THIS CHAPTER WE WILL consider the simultaneous solution to a set of linear equations. We will use Cramer's rule, Gauss elimination, Gauss-Jordan elimination, and the Gauss-Seidel method. In addition, we will study the problem of *ill conditioning* by generating a set of Hilbert matrices. We will develop special FORTRAN application programs. These include the solution to a set of equations with multiple constant vectors and equations with complex coefficients. We will also present a program for producing the best fit to an overdetermined system.

We will begin by describing linear equations and providing a simple example of simultaneous equations.

LINEAR EQUATIONS AND SIMULTANEOUS SOLUTIONS

A *linear equation* consists of a sum of terms such as:

$$Ax + By + Cz = D$$

In this equation, x, y and z are variables and A, B, C, and D are constants. No more than one variable can occur in each term, and it must be present to the first power. Thus, expressions such as:

$$2x^2 + 3y^2 = 4$$
$$\sin(x) + \log(y) = 9$$

and

$$xy = 2$$

are *nonlinear equations* if the variables are x and y.

An equation such as:

$$\frac{x}{A} + y \log (B) = p$$

is linear in the parameters x and y. On the other hand, if the symbols A and B are considered as the variables, and the symbols x, y and p are known values, then the equation is nonlinear.

Linear equations occur frequently in all branches of science and engineering, so effective methods for their solution are necessary. For the particular case where there are several unknowns and an equal number of independent equations, a unique solution is possible. In this chapter, we will consider several different methods for the solution of a set of linear equations. The solution of nonlinear equations is discussed in other chapters.

The two linear equations:

$$x - 2y = 1$$
$$2x + y = 7$$

represent two straight lines in the x-y plane. The simultaneous solution of these two equations represents the intersection of the two lines. Therefore, a graphical solution can be obtained by plotting the lines and finding the point of intersection.

The first equation has a y-intercept of -0.5 and a slope of 0.5. This can be seen by rearranging the equation as:

$$y = -0.5 + 0.5x$$

The second equation has an intercept of 7 and a slope of -2:

$$y = 7 - 2x$$

The intersection of the two lines occurs at the location:

$$x = 3, \quad y = 1$$

which represents the solution to the problem. This solution can be verified by substitution into the original equations.

Another method for finding the simultaneous solution to these two equations is to multiply the second equation by 2 and add it to the first equation. The resulting equation contains only the variable x, and so it can be solved directly.

$$
\begin{aligned}
x - 2y &= 1 \\
4x + 2y &= 14 \\
\hline
5x &= 15 \quad \text{or} \quad x = 3
\end{aligned}
$$

This value of x can then be substituted into either one of the original equations to find the corresponding value of y. Both of the above methods are suitable for solving two simultaneous equations. But, in general, these methods are tedious for larger numbers of equations. In the next section we will study a more sophisticated method of solving simultaneous equations, using matrices and vectors.

SOLUTION BY CRAMER'S RULE

A technique known as Cramer's rule is useful for solving two or three simultaneous equations. This is a particularly powerful method when the solution is performed by hand, or when a pocket calculator is used. In this approach, the equations are written with the unknowns on one side of the equal sign, and the constant terms on the other. Terms containing the same unknowns are vertically aligned. The two equations in the previous section were initially written in this form.

The coefficients of the unknowns are placed into a matrix known as the *coefficient matrix*. The constant terms are put into a separate vector. For our two equations, this produces the matrix A and the vector $\mathbf{z}$.

$$A = \begin{bmatrix} 1 & -2 \\ 2 & 1 \end{bmatrix} \qquad \mathbf{z} = \begin{bmatrix} 1 \\ 7 \end{bmatrix}$$

The solution is found from the relationship:

$$x = \frac{D_1}{D} \qquad y = \frac{D_2}{D}$$

where D is the determinant of the coefficient matrix:

$$D = \begin{vmatrix} 1 & -2 \\ 2 & 1 \end{vmatrix} = (1)(1) - (2)(-2) = 5$$

The determinant D_1 is found by substituting the constant vector into column 1 of the matrix. This column corresponds to the unknown x:

$$D_1 = \begin{vmatrix} 1 & -2 \\ 7 & 1 \end{vmatrix} = (1)(1) - (7)(-2) = 15$$

Similarly, D_2 is obtained by substituting the constant vector into column 2:

$$D_2 = \begin{vmatrix} 1 & 1 \\ 2 & 7 \end{vmatrix} = (1)(7) - (2)(1) = 5$$

The solution is then:

$$x = \frac{15}{5} = 3 \qquad \text{and} \qquad y = \frac{5}{5} = 1$$

Of course, some sets of equations have no unique solutions. We will now examine how Cramer's rule handles such equations.

Linear Dependence

We can see that if the determinant of the coefficient matrix is zero, then no unique solution is possible. This will occur whenever there is a *linear dependence* among the equations; that is, when one equation can be obtained from a combination of the others. Here, the matrix is singular. As an example of linear dependence, consider the two equations:

$$x + y = 5$$
$$2x + 2y = 2$$

The coefficient matrix for these two equations is:

$$\begin{bmatrix} 1 & 1 \\ 2 & 2 \end{bmatrix}$$

and the corresponding determinant is zero. These two equations represent parallel lines, which do not, of course, intersect. Consequently, there can be no solution.

As a second example, consider the two equations:

$$x + y = 5$$
$$2x + 2y = 10$$

This example produces the same singular coefficient matrix as the previous one. Here, however, the two lines lie on top of each other, and so there are an infinite number of solutions.

Now that we have a powerful method of solving two or three simultaneous equations, let us consider a practical application of the method. We will solve the problem of the following section using a FORTRAN program we developed in Chapter 3.

Example: A Direct-Current Electrical Circuit

An interesting practical example where simultaneous equations must be solved is provided by an electrical circuit. Consider, for example, the network of resistors and direct-current voltage sources shown in Figure 4.1. This circuit contains four nodes and six branches. There is a 20-volt source on the left side of the network, and a 5-volt source on the right side. In addition, there are six resistors of known value.

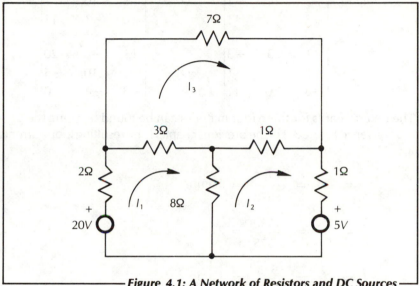

Figure 4.1: A Network of Resistors and DC Sources

The problem is to find the resulting branch currents and the voltages across each resistor. The electrical currents in the six separate branches can be determined by solving six simultaneous equations. The program can be simplified, however, by considering three loop currents and the corresponding three simultaneous equations. Then the branch currents can be found from the loop currents.

The loop current in the lower-left loop is designated as I_1, the loop current in the lower-right loop is I_2 and the loop current in the upper loop is I_3. Then the three loop equations can be derived from the Kirchhoff voltage law by going around each loop in a counter-clockwise direction:

$$13I_1 - 8I_2 - 3I_3 - 20 = 0 \text{ (lower-left loop)}$$
$$-8I_1 + 10I_2 - I_3 + 5 = 0 \text{ (lower-right loop)}$$
$$-3I_1 - I_2 + 11I_3 = 0 \text{ (upper loop)}$$

The corresponding coefficient matrix and constant vector are:

$$\begin{bmatrix} 13 & -8 & -3 \\ -8 & 10 & -1 \\ -3 & -1 & 11 \end{bmatrix} \quad \begin{bmatrix} 20 \\ -5 \\ 0 \end{bmatrix}$$

Proceeding by Cramer's rule, we find:

$$D = \begin{vmatrix} 13 & -8 & -3 \\ -8 & 10 & -1 \\ -3 & -1 & 11 \end{vmatrix} \qquad D_1 = \begin{vmatrix} 20 & -8 & -3 \\ -5 & 10 & -1 \\ 0 & -1 & 11 \end{vmatrix}$$

$$D_2 = \begin{vmatrix} 13 & 20 & -3 \\ -8 & -5 & -1 \\ -3 & 0 & 11 \end{vmatrix} \qquad D_3 = \begin{vmatrix} 13 & -8 & 20 \\ -8 & 10 & -5 \\ -3 & -1 & 0 \end{vmatrix}$$

The determinants for these four matrices can be found by using the program given in Figure 3.3 of the previous chapter. The resulting loop currents are:

$$I_1 = \frac{D_1}{D} = \frac{1725}{575} = 3 \text{ amps}$$

$$I_2 = \frac{D_2}{D} = \frac{1150}{575} = 2 \text{ amps}$$

$$I_3 = \frac{D_3}{D} = \frac{575}{575} = 1 \text{ amp}$$

While the above approach gave us the correct answer, it was unnecessarily complicated. Nine separate numbers were entered into the computer program for each of the four determinants. This required the entry of thirty-six separate values.

We will now study an efficient program designed specifically to use Cramer's rule to solve simultaneous equations.

FORTRAN PROGRAM: A MORE ELEGANT USE OF CRAMER'S RULE

The program shown in Figure 4.3 simplifies the process considerably. The mathematical operations are exactly the same as in our first solution, but with this version, it is only necessary to enter twelve numbers: nine for the coefficients and three for the constant vector.

Running the Program

Type up the program shown in Figure 4.3 and execute it. The coefficients and constant term for the first equation are entered in order. The return key is pressed after each number is given. The corresponding values for the second equation are entered next and then the data for the third equation are entered. The input dialogue and the solution will look like Figure 4.2.

```
      Simultaneous  solution  by  Cramer's  rule
      Equation    1
          1:  13
          2:  -8
          3:  -3
       C:  20

      Equation    2
          1:  -8
          2:  10
          3:  -1
       C:  -5

      Equation    3
          1:  -3
          2:  -1
          3:  11
       C:  0


   1.3000E+01  -8.0000E+00  -3.0000E+00   2.0000E+01
  -8.0000E+00   1.0000E+01  -1.0000E+00  -5.0000E+00
  -3.0000E+00  -1.0000E+00   1.1000E+01   0.0000E+00


      Solution
   3.0000E+00   2.0000E+00   1.0000E+00

   More?  N
```

Figure 4.2:
Input/Output Dialogue: Solving Simultaneous Equations by Cramer's Rule

The program will begin again. Another equation can be solved at this point.

```
C       PROGRAM SIMQ1
C
C -- FORTRAN program to solve three
C -- simultaneous equations by Cramer's rule.
C -- Apr 28, 81
C
        LOGICAL ERROR
        INTEGER LENGTH, RMAX, CMAX, YESNO, IN, OUT
        REAL A(3,3), Y(3), COEF(3)
        COMMON /INOUT/ IN, OUT
C
        IN = 1
        OUT = 1
10      WRITE(OUT, 101)
        CALL INPUT(A, Y, LENGTH)
        CALL CRAMER(A, Y, COEF, ERROR)
        IF (.NOT. ERROR) CALL OUTPUT(A, Y, COEF, LENGTH)
```

Figure 4.3: Solution to Three Linear Equations by Cramer's Rule

```
        WRITE(OUT, 102)
        READ(IN, 103) YESNO
        IF (YESNO .EQ. 'Y' .OR. YESNO .EQ. 'y') GOTO 10
        STOP
101     FORMAT('1 Simultaneous solution by Cramer''s rule')
102     FORMAT(/' More? ')
103     FORMAT(A1)
        END
        SUBROUTINE INPUT(A, Y, N)
C -- Get values for N and arrays A and Y.
C
        INTEGER N, IN, OUT, I, J
        REAL A(N,N), Y(N)
        COMMON /INOUT/ IN, OUT
C
        N = 3
        DO 20 I = 1, N
          WRITE(OUT, 101) I
          DO 10 J = 1, N
            WRITE(OUT, 102) J
            READ(IN, 103) A(I,J)
10        CONTINUE
          WRITE(OUT, 104)
          READ(IN, 103) Y(I)
20      CONTINUE
        RETURN
101     FORMAT(' Equation ', I3/)
102     FORMAT('+',I4, ': ')
103     FORMAT(F10.0)
104     FORMAT('+ C: ')
        END
        SUBROUTINE CRAMER(A, Y, COEF, ERROR)
C
C -- Solve 3 simultaneous equations by Cramer's rule.
C
        LOGICAL ERROR
        INTEGER N, I, J, IN, OUT
        REAL A(3,3), Y(3), COEF(3), B(3,3), DET
        COMMON /INOUT/ IN, OUT
C
        ERROR = .FALSE.
        N = 3
        DO 20 I = 1, N
          DO 10 J = 1, N
            B(I,J) = A(I,J)
10        CONTINUE
20      CONTINUE
        DET = DETERM(B)
        IF (DET .EQ. 0.0) GOTO 90
        CALL SETUP(DET, A, B, Y, COEF,1)
        CALL SETUP(DET, A, B, Y, COEF,2)
        CALL SETUP(DET, A, B, Y, COEF,3)
        RETURN
```

Figure 4.3: Solution to Three Linear Equations by Cramer's Rule (cont.)

```
C
90      ERROR = .TRUE.
        WRITE(OUT, 101)
        RETURN
101     FORMAT(' ERROR--matrix singular')
        END
        FUNCTION DETERM(X)
C
C -- Find the determinant of the 3-by-3 matrix X.
C
        REAL X(3,3), SUM
C
        SUM = X(1,1) * (X(2,2)*X(3,3) - X(3,2)*X(2,3))
        SUM = SUM - X(1,2) * (X(2,1)*X(3,3) - X(3,1)*X(2,3))
        SUM = SUM + X(1,3) * (X(2,1)*X(3,2) - X(3,1)*X(2,2))
        DETERM = SUM
        RETURN
        END
        SUBROUTINE SETUP(DET, A, B, Y, COEF, J)
C
C -- Set up the numerator matrices.
C
        INTEGER I, J
        REAL A(3,3), B(3,3), Y(3), COEF(3), DET
C
        DO 10 I = 1, 3
          B(I,J) = Y(I)
          IF (J .GT. 1) B(I, J-1) = A(I, J-1)
10      CONTINUE
        COEF(J) = DETERM(B) / DET
        RETURN
        END
        SUBROUTINE OUTPUT(A, Y, COEF, N)
C
C -- Print out the answers.
C
        INTEGER N, IN, OUT, I, J
        REAL A(N,N), Y(N), COEF(N)
        COMMON /INOUT/ IN, OUT
C
        WRITE(OUT, 101) ((A(I,J), J = 1, N), Y(I), I = 1, N)
        WRITE(OUT, 102)
        WRITE(OUT, 101) (COEF(I), I = 1, N)
        RETURN
101     FORMAT(1P4E12.4)
102     FORMAT(/ '   Solution')
        END
```

Figure 4.3: Solution to Three Linear Equations Cramer's Rule (cont.)

SOLUTION BY GAUSS ELIMINATION

Cramer's rule is an effective method for solving two or three simultaneous equations, particularly when the solution is performed by hand. However, the computation time increases with the fourth power of the matrix size. Consequently, it will take about 16 times longer to solve six simultaneous equations than it does to solve three equations. The Gauss elimination method is more efficient. Computation time increases with the third power of the number of equations, and so it will take about 8 times longer to solve six equations than it does to solve three.

The Gauss elimination method can be readily programmed on a digital computer. However, the technique is fairly complicated, and therefore is not suitable for hand solution. Furthermore, the operations involve frequent multiplication, division and subtraction. The consequent loss of precision places a practical upper limit on the number of equations that can be solved simultaneously. Let us go through the method step by step.

The Steps of the Gauss Method

With the Gauss elimination method, the original equations are manipulated so that the coefficient matrix contains a value of unity at each point on the major diagonal and zero at each position below and to the left of the major diagonal.

Two basic types of matrix operations are used in the Gauss elimination method: scalar multiplication and addition. Any equation can be multiplied by a constant without changing the result. This is equivalent to multiplying one row of the coefficient matrix and the corresponding element in the constant vector by the same value. Also, any equation can be replaced by the sum of two equations.

The following steps will demonstrate the use of Gauss elimination using the equations we derived from the electric circuit shown in Figure 4.1. Initially, the coefficient matrix and the constant vector are:

$$\begin{bmatrix} 13 & -8 & -3 \\ -8 & 10 & -1 \\ -3 & -1 & 11 \end{bmatrix} \qquad \begin{bmatrix} 20 \\ -5 \\ 0 \end{bmatrix}$$

The first variable is eliminated from all but the first equation. The equations are manipulated to produce the value of unity at the top of the first column. The remaining positions of the column become zeros. The operation is performed in the following way. The entire first row is divided by the first element in the row (the pivot element). This generates the value of unity in the first diagonal position. By this means, the first equation becomes:

$$\begin{bmatrix} 1 & -0.61 & -0.23 \end{bmatrix} \qquad \begin{bmatrix} 1.5 \end{bmatrix}$$

The first unknown is eliminated from the second row by combining the first two rows. The new first row is multiplied by the first element in the second row, then subtracted from the second row. The new second row is:

$$\begin{bmatrix} 0 & 5.1 & -2.8 \end{bmatrix} \qquad \begin{bmatrix} 7.3 \end{bmatrix}$$

In a similar fashion, the first variable is eliminated from the third equation. The three equations are now approximately:

$$\begin{bmatrix} 1 & -0.61 & -0.23 \\ 0 & 5.1 & -2.8 \\ 0 & -2.8 & 10.3 \end{bmatrix} \qquad \begin{bmatrix} 1.5 \\ 7.3 \\ 4.6 \end{bmatrix}$$

The next step is to produce the value of unity in the second position of the second row (the new pivot). The second line is divided by the second element. The second variable is eliminated from the third equation by generating a zero in the second position, just under the pivot element. The three equations now look like:

$$\begin{bmatrix} 1 & -0.61 & -0.23 \\ 0 & 1 & -0.56 \\ 0 & 0 & 8.7 \end{bmatrix} \qquad \begin{bmatrix} 1.5 \\ 1.4 \\ 8.7 \end{bmatrix}$$

The final step of this phase is to obtain a value of unity in the third pivot position. This is accomplished by dividing the third equation by the pivot value to give:

$$\begin{bmatrix} 1 & -0.61 & -0.23 \\ 0 & 1 & -0.56 \\ 0 & 0 & 1 \end{bmatrix} \qquad \begin{bmatrix} 1.5 \\ 1.4 \\ 1 \end{bmatrix}$$

The result corresponds to the three equations:

$$x - 0.61y - 0.23z = 1.5$$
$$y - 0.56z = 1.4$$
$$z = 1$$

The third equation can be solved directly since it has only one unknown. The result is:

$$z = 1$$

The second equation becomes:

$$y = 1.4 + 0.56z$$

By substituting the value of z into this equation, we find the value of y to be 2. Substituting the values of y and z into the first equation produces the value of 3 for x. This phase of the calculations is known as *back substitution*.

Improving the Accuracy of the Gauss Method

The accuracy of the Gauss elimination method can be improved by interchanging two rows so that the element with the largest absolute magnitude becomes the pivot element. Suppose, for example, that the previous three equations had been originally written in a different order:

$$\begin{bmatrix} -3 & -1 & 11 \\ 13 & -8 & -3 \\ -8 & 10 & -1 \end{bmatrix} \quad \begin{bmatrix} 0 \\ 20 \\ -5 \end{bmatrix}$$

Then the top equation could be divided by -3 to put the value of unity in the pivot position. However, the result will be more accurate if the first and second rows are interchanged first. This will put the larger value of 13 in the pivot position. After the first variable is eliminated from the second and third equations, the result is:

$$\begin{bmatrix} 1 & -0.61 & -0.23 \\ 0 & -2.8 & 10.3 \\ 0 & 5.1 & -2.8 \end{bmatrix} \quad \begin{bmatrix} 1.5 \\ 4.6 \\ 7.3 \end{bmatrix}$$

Again, it would be best to interchange the second and third equations to put the larger element of 5.1 in the second pivot position.

There is another reason why rows may have to be interchanged. If a zero element appears on the major diagonal, then it will not be possible to divide the row by this pivot element. But interchanging this row with one that is below it will remove the zero from the pivot position.

Now that we have gone through the rather tedious process of using Gaussian elimination to solve equations by hand, we can appreciate the elegance of the FORTRAN program in the following section.

FORTRAN PROGRAM: THE GAUSS ELIMINATION METHOD

The program shown in Figure 4.4 can be used to simultaneously solve a set of linear equations by Gaussian elimination. The program is written to handle up to eight equations. The size can be increased by changing the two variables called MAXR and MAXC and the corresponding dimension statements. Type up the program and execute it.

```
C      PROGRAM SIMQ2
C
C -- FORTRAN program to solve
C -- simultaneous equations by Gauss elimination.
C -- Apr 30, 81
C
```

Figure 4.4: Simultaneous Solution by Gauss Elimination

```
          LOGICAL ERROR
          INTEGER LENGTH, MAXR, MAXC, IN, OUT
          REAL A(8,8), Y(8), COEF(8)
          COMMON /INOUT/ IN, OUT, MAXR, MAXC
C
          IN = 1
          OUT = 1
          MAXR = 8
          MAXC = 8
          WRITE(OUT, 101)
10        CALL INPUT(A, Y, LENGTH)
          IF (LENGTH .LT. 2) STOP
          CALL GAUSS(A, Y, COEF, LENGTH, ERROR)
          IF (.NOT. ERROR) CALL OUTPUT(A, Y, COEF, LENGTH)
          GOTO 10
101       FORMAT('1 Simultaneous solution by Gauss elimination')
          END
          SUBROUTINE INPUT(A, Y, N)
C -- Get values for N and arrays A and Y.
C
          INTEGER N, IN, OUT, I, J, M
          REAL A(8,8), Y(8)
          COMMON /INOUT/ IN, OUT, MAXR
C
5         WRITE(OUT, 105)
          READ(IN, 106) N
          M = N
          IF (N .GT. MAXR) GOTO 5
          IF (N .LT. 2) RETURN
          DO 20 I = 1, N
            WRITE(OUT, 101) I
            DO 10 J = 1, N
              WRITE(OUT, 102) J
              READ(IN, 103) A(I,J)
10          CONTINUE
            WRITE(OUT, 104)
            READ(IN, 103) Y(I)
20        CONTINUE
          RETURN
101       FORMAT(' Equation ', I3/)
102       FORMAT('+',I4, ': ')
103       FORMAT(F10.0)
104       FORMAT('+ C: ')
105       FORMAT(/' How many equations? ')
106       FORMAT(I2)
          END
          SUBROUTINE GAUSS(A, Y, COEF, NCOL, ERROR)
C
C -- Simultaneous solution by Gauss elimination.
C -- Apr 26, 81
C
```

Figure 4 .4: Simultaneous Solution by Gauss Elimination (cont.)

```
          LOGICAL ERROR
          INTEGER NCOL, I, J, IN, OUT, N, N1, I1, K, L
          REAL A(8,8), Y(8), COEF(8)
          REAL B(8,8), W(8), BIG, AB, SUM, T
          COMMON /INOUT/ IN, OUT
C
          ERROR = .FALSE.
          N = NCOL
          DO 20 I = 1, N
            DO 10 J = 1, N
              B(I,J) = A(I,J)
10          CONTINUE
            W(I) = Y(I)
20        CONTINUE
          N1 = N - 1
          DO 70 I = 1, N1
            BIG = ABS(B(I,I))
            L = I
            I1 = I + 1
            DO 30 J = I1, N
              AB = ABS(B(J,I))
              IF (AB .LE. BIG) GOTO 30
                BIG = AB
                L = J
30          CONTINUE
            IF (BIG .EQ. 0.0) GOTO 70
            IF (L .EQ. I) GOTO 50
C -- Interchange rows to put largest element on diagonal.
            DO 40 J = 1, N
              CALL SWAP(B(L,J), B(I,J))
40          CONTINUE
            CALL SWAP(W(I), W(L))
50          CONTINUE
            DO 60 J = I1, N
              T = B(J,I) / B(I,I)
              DO 55 K = I1, N
                B(J,K) = B(J,K) - T * B(I,K)
55            CONTINUE
              W(J) = W(J) - T * W(I)
60          CONTINUE
70        CONTINUE
          IF (B(N,N) .EQ. 0.0) GOTO 99
          COEF(N) = W(N) / B(N,N)
          I = N - 1
C -- Back substitution.
80        SUM = 0
            I1 = I + 1
            DO 75 J = I1, N
              SUM = SUM + B(I,J) * COEF(J)
75          CONTINUE
```

Figure 4.4: Simultaneous Solution by Gauss Elimination (cont.)

```
              COEF(I) = (W(I) - SUM) / B(I,I)
                I = I - 1
              IF (I .GT. 0) GOTO 80
              RETURN
  99          WRITE(OUT, 999)
              RETURN
  999         FORMAT(' ERROR--matrix singular')
              END
              SUBROUTINE SWAP(A,B)
C --  Interchange two values.
              REAL A, B, HOLD
C
              HOLD = A
              A = B
              B = HOLD
              RETURN
              END
              SUBROUTINE OUTPUT(A, Y, COEF, N)
C
C --  Print out the answers.
C
              INTEGER N, IN, OUT, I, J
              REAL A(8,8), Y(8), COEF(8)
              COMMON /INOUT/ IN, OUT
C
              DO 10 I = 1, N
                WRITE(OUT, 101) (A(I,J), J = 1, N), Y(I)
  10          CONTINUE
              WRITE(OUT, 102)
              WRITE(OUT, 101) (COEF(I), I = 1, N)
              RETURN
  101         FORMAT(1P6E12.4)
  102         FORMAT(/ '    Solution')
              END
```

Figure 4.4: Simultaneous Solution by Gauss Elimination (cont.)

Running the Program

This program is similar to the previous one, but it begins by asking for the number of equations. Answer this question with a number from 2 to 8, then press the carriage return. Enter the coefficients for each equation and the corresponding constant vectors in turn. Press the carriage return after each number is entered. Enter the values for the set of equations we considered earlier in this chapter and verify that the solution vector is [3 2 1]. The results should look like Figure 4.5.

```
       Simultaneous solution by Gauss elimination

   How many equations?
   3
   Equation    1
       1: 13
       2: -8
       3: -3
    C: 20

   Equation    2
       1: -8
       2: 10
       3: -1
    C: -5

   Equation    3
       1: -3
       2: -1
       3: 11
    C: 0

    1.3000E+01  -8.0000E+00  -3.0000E+00   2.0000E+01
   -8.0000E+00   1.0000E+01  -1.0000E+00  -5.0000E+00
   -3.0000E+00  -1.0000E+00   1.1000E+01   0.0000E+00

      Solution
    3.0000E+00   2.0000E+00   1.0000E+00

   How many equations?
   0
```

**Figure 4.5: Input/Output Dialogue:
Solution of the Electric Circuit Problem Using Gauss Elimination**

At the completion of the task, the program begins again. This time find the simultaneous solution to the following three equations and write down the answer for later reference.

$$\begin{bmatrix} 1 & 1 & 1 \\ 2 & 1 & -1 \\ 3 & 1 & -3 \end{bmatrix} \quad \begin{bmatrix} 6 \\ 1 \\ -4 \end{bmatrix}$$

We will come back to this set at the end of the next section.

For the third task, enter two lines that are exactly the same, for example:

$$\begin{bmatrix} 1 & 1 & 1 \\ 1 & 1 & 1 \\ 2 & 1 & -1 \end{bmatrix} \quad \begin{bmatrix} 6 \\ 6 \\ 1 \end{bmatrix}$$

The program should print an error message indicating that the matrix is singular. The program can be aborted by entering a zero or negative value for the number of equations.

SOLUTION BY GAUSS-JORDAN ELIMINATION

A variation of the Gauss elimination method is known as the Gauss-Jordan method. This approach shares most of the advantages and disadvantages of the Gauss elimination technique. Execution time is a third-order function of the matrix size, and there are many multiplication, division, and subtraction operations that contribute to loss of accuracy. Furthermore, the Gauss-Jordan algorithm is more complicated than the one for Gauss elimination. Nevertheless, the Gauss-Jordan technique will generally be the most useful of all. In fact, we will use it in later chapters of this book. Its usefulness lies in the fact that the inverse of the coefficient matrix is readily obtained along with the solution vector. Let us outline the steps of this method.

Details of the Gauss-Jordan Method

In the Gauss-Jordan method, the elements of the major diagonal are converted to unity, as they are in the Gauss method. But now the elements both above and below the major diagonal are converted to zeros. Thus, the coefficient matrix is converted to a unit matrix. The resulting constant vector then becomes the solution vector. For the Gauss-Jordan solution of the electrical circuit given in Figure 4.1, the final set of values becomes:

$$\begin{bmatrix} 1 & 0 & 0 \\ 0 & 1 & 0 \\ 0 & 0 & 1 \end{bmatrix} \quad \begin{bmatrix} 3 \\ 2 \\ 1 \end{bmatrix}$$

This corresponds to the three equations:

$$x \qquad\qquad = 3$$
$$\qquad y \qquad = 2$$
$$\qquad\qquad z = 1$$

for which the solution is:

$$x = 3, \quad y = 2, \quad z = 1$$

Suppose that a unit matrix is initially placed to the right of the original set of equations. Then, if all the operations are performed on this matrix as they are performed on the other elements, this unit matrix will be converted into the inverse of the original coefficient matrix at the conclusion of the

```
C       PROGRAM SIMQ3
C
C -- FORTRAN program to solve simultaneous equations
C -- by Gauss-Jordan elimination.
C -- Subroutines GAUSSJ and SWAP are also needed.
C -- May  4, 81
C
        LOGICAL ERROR
        INTEGER SIZE, MAXR, MAXC, IN, OUT, INDEX(8,3), NVEC
        REAL A(8,8), Y(8), COEF(8), B(8,8)
        COMMON /INOUT/ IN, OUT, MAXR, MAXC, ERROR
        DATA NVEC/1/
C
        IN = 1
        OUT = 1
        MAXR = 8
        MAXC = 8
        WRITE(OUT, 101)
10      CALL INPUT(A, Y, SIZE)
        IF (SIZE .LT. 2) GOTO 100
        DO 30 I = 1, SIZE
           DO 20 J = 1, SIZE
              B(I,J) = A(I,J)
20         CONTINUE
           COEF(I) = Y(I)
30      CONTINUE
        CALL GAUSSJ(B, COEF, INDEX, SIZE, MAXC, NVEC, ERROR, OUT)
        IF (.NOT. ERROR) CALL OUTPUT(A, Y, COEF, SIZE)
        GOTO 10
100     STOP
101     FORMAT('1 Simultaneous solution by',
     *  ' Gauss-Jordan elimination')
        END
        SUBROUTINE INPUT(A, Y, N)
C -- Get values for N and arrays A and Y.
C
        INTEGER N, IN, OUT, I, J, M, MAXR
        REAL A(8,8), Y(8)
        COMMON /INOUT/ IN, OUT, MAXR
C
5       WRITE(OUT, 105)
        READ(IN, 106) N
        M = N
        IF (N .GT. MAXR) GOTO 5
        IF (N .LT. 2) RETURN
        DO 20 I = 1, N
           WRITE(OUT, 101) I
           DO 10 J = 1, N
              WRITE(OUT, 102) J
              READ(IN, 103) A(I,J)
10         CONTINUE
           WRITE(OUT, 104)
           READ(IN, 103) Y(I)
20      CONTINUE
```

Figure 4.6: Solution of Simultaneous Equations by Gauss-Jordan Elimination

```
          RETURN
101       FORMAT(' Equation ', I3/)
102       FORMAT('+',I4, ': ')
103       FORMAT(F10.0)
104       FORMAT('+ C: ')
105       FORMAT(/' How many equations? ')
106       FORMAT(I2)
          END
          SUBROUTINE OUTPUT(A, Y, COEF, N)
C
C -- Print out the answers.
C
          LOGICAL ERROR
          INTEGER N, IN, OUT, I, J, MAXR, MAXC
          REAL A(8,8), Y(8), COEF(8)
          COMMON /INOUT/ IN, OUT, MAXR, MAXC, ERROR
C
          DO 10 I = 1, N
            WRITE(OUT, 101) (A(I,J), J = 1, N), Y(I)
10        CONTINUE
          WRITE(OUT, 102)
          IF (ERROR) RETURN
          WRITE(OUT, 101) (COEF(I), I = 1, N)
          RETURN
101       FORMAT(1P6E12.4)
102       FORMAT(/ '    Solution')
          END
```

Figure 4.6:
Solution of Simultaneous Equations by Gauss-Jordan Elimination (cont.)

```
          SUBROUTINE GAUSSJ(B, W, INDEX, NROW, MAX, NVEC, ERROR, LUN)
C
C -- Simultaneous solution by the Gauss-Jordan method.
C -- Subroutine SWAP is needed.
C
C -- B ---- Coefficient matrix becomes inverse matrix.
C -- W ---- Constant vector/matrix becomes solution.
C -- INDEX- Work array, needs three columns.
C -- NROW - Current matrix size.
C -- MAX  - Dimensioned matrix size.
C -- NVEC - Number of constant vectors (columns).
C -- ERROR- Logical error flag. .TRUE. if matrix singular.
C -- LUN  - Logical unit number for error message.
C -- May  3, 81
C
          LOGICAL ERROR
          INTEGER NROW, I, J, LUN, K, L, NVEC
          INTEGER IROW, ICOL, L1, INDEX(MAX,3)
          REAL B(MAX,1), W(MAX,1), BIG, SUM, T, PIVOT, DETERM
C
          ERROR = .FALSE.
```

Figure 4.7: Gauss-Jordan Subroutine

```
         DO 10 I = 1, NROW
           INDEX(I,3) = 0
10       CONTINUE
         DETERM = 1.0
         DO 90 I = 1, NROW
C -- Search for largest element.
         BIG = 0.0
           DO 20 J = 1, NROW
            IF (INDEX(J,3) .EQ. 1) GOTO 20
              DO 15 K = 1, NROW
                IF (INDEX(K,3) .GT. 1) GOTO 199
                IF (INDEX(K,3) .EQ. 1) GOTO 15
                IF (ABS(B(J,K)) .LE. BIG) GOTO 15
                  IROW = J
                  ICOL = K
                  BIG = ABS(B(J,K))
15            CONTINUE
20          CONTINUE
         INDEX(ICOL,3) = INDEX(ICOL,3) + 1
         INDEX(I,1) = IROW
         INDEX(I,2) = ICOL
C -- Interchange rows to put pivot on diagonal.
         IF (IROW .EQ. ICOL) GOTO 40
           DETERM = -DETERM
           DO 25 L = 1, NROW
             CALL SWAP(B(IROW,L), B(ICOL,L))
25         CONTINUE
           IF (NVEC .EQ. 0) GOTO 40
           DO 30 L = 1, NVEC
             CALL SWAP(W(IROW,L), W(ICOL, L))
30         CONTINUE
C -- Divide pivot row by pivot element.
40       PIVOT = B(ICOL, ICOL)
         DETERM = DETERM * PIVOT
         B(ICOL, ICOL) = 1.0
         DO 45 L = 1, NROW
           B(ICOL, L) = B(ICOL, L) / PIVOT
45       CONTINUE
         IF (NVEC .EQ. 0) GOTO 60
           DO 50 L = 1, NVEC
             W(ICOL, L) = W(ICOL, L) / PIVOT
50         CONTINUE
C -- Reduce nonpivot terms.
60       DO 80 L1 = 1, NROW
           IF (L1 .EQ. ICOL) GOTO 80
             T = B(L1, ICOL)
             B(L1, ICOL) = 0.0
             DO 65 L = 1, NROW
               B(L1, L) = B(L1, L) - B(ICOL, L) * T
65           CONTINUE
             IF (NVEC .EQ. 0) GOTO 80
               DO 70 L = 1, NVEC
                 W(L1, L) = W(L1, L) - W(ICOL, L) * T
```

Figure 4.7: Gauss-Jordan Subroutine (cont.)

```
70                  CONTINUE
80           CONTINUE
90       CONTINUE
C -- End of I loop.
C -- Interchange columns.
         DO 120 I = 1, NROW
            L = NROW - I + 1
            IF (INDEX(L,1) .EQ. INDEX(L,2)) GOTO 120
               IROW = INDEX(L,1)
               ICOL = INDEX(L,2)
               DO 110 K = 1, NROW
                  CALL SWAP(B(K, IROW), B(K, ICOL))
110            CONTINUE
120      CONTINUE
         DO 130 K = 1, NROW
            IF (INDEX(K, 3) .NE. 1) GOTO 199
130      CONTINUE
         RETURN
199      WRITE(LUN, 999)
         ERROR = .TRUE.
         RETURN
999      FORMAT(' ERROR--matrix singular')
         END
         SUBROUTINE SWAP(A,B)
C -- Interchange two values.
         REAL A, B, HOLD
C
         HOLD = A
         A = B
         B = HOLD
         RETURN
         END
```

Figure 4.7: Gauss-Jordan Subroutine (cont.)

Type up the program and execute it. Try out the electrical circuit equations and verify that the solution is [3 2 1]. Then try the set:

$$\begin{bmatrix} 1 & 1 & 1 \\ 2 & 1 & -1 \\ 3 & 1 & -3 \end{bmatrix} \qquad \begin{bmatrix} 6 \\ 1 \\ -4 \end{bmatrix}$$

Notice that the answer, for this example, is not the same as the one found by the Gauss method. Furthermore, the set:

$$x = 1, \quad y = 2, \quad z = 3$$

is also a solution. How can there be more than one solution to a set of linear equations? The answer is that one equation is a linear combination of the other two. The third equation, in this example, is equal to twice the second equation minus the first equation. Thus, the coefficient matrix is singular. Actually, if the above three equations are given to the Cramer's rule program at the beginning of this chapter, the singularity will be readily found.

Unfortunately, this kind of linear dependence is difficult to find. The determinant of the coefficient matrix may be small, but it does not equal zero, because of roundoff errors that accumulate during the elimination process. Fortunately, different algorithms for solving simultaneous equations will usually give different answers when there is linear dependence. Consequently, if there is any question of linear dependence, the equations should be solved by at least two different methods.

We have seen how the Gauss-Jordan elimination program works for one coefficient matrix and its corresponding constant vector. Now we will explore ways of using and refining this program to deal conveniently with *multiple* constant vectors. To illustrate this situation we will return to the electrical circuit example that we set up earlier in this chapter. We will also continue our discussion of inverse coefficient matrices and their use in solving simultaneous equations.

MULTIPLE CONSTANT VECTORS AND MATRIX INVERSION

In the previous chapter, we mentioned that the solution to a set of linear equations could be obtained by multiplying the inverse of the coefficient matrix by the constant vector. For example, if the coefficient matrix is A and the constant vector is $\mathbf{y}$, then the solution vector $\mathbf{b}$ is:

$$\mathbf{b} = A^{-1}\ \mathbf{y}$$

But the methods developed in this chapter (Cramer's rule, Gauss elimination and Gauss-Jordan elimination) obtain the solution to a set of linear equations by direct methods. The inverse of the coefficient matrix is not utilized.

Sometimes we need to solve several sets of simultaneous equations that all have the same coefficient matrix but different constant vectors. One possible approach is to invert the coefficient matrix. The separate solution vectors can then be obtained from the product of the inverted matrix and each constant vector. Even in this example, however, it will generally be faster to perform a Gauss-Jordan elimination on the matrix and all the constant vectors simultaneously. In fact, the Gauss-Jordan procedure given in the previous section is actually programmed for multiple constant vectors.

Consider, for example, the electrical circuit shown in Figure 4.1. Suppose that we would like to determine the loop currents for three different circuit configurations:

1. The original circuit.
2. The circuit with the 5-volt source reversed.
3. The circuit with both voltage sources reversed.

These three different configurations correspond to the following three sets of equations:

$$
\begin{bmatrix} 13 & -8 & -3 \\ -8 & 10 & -1 \\ -3 & -1 & 11 \end{bmatrix}
\begin{bmatrix} 20 \\ -5 \\ 0 \end{bmatrix}
$$

$$
\begin{bmatrix} 13 & -8 & -3 \\ -8 & 10 & -1 \\ -3 & -1 & 11 \end{bmatrix}
\begin{bmatrix} 20 \\ 5 \\ 0 \end{bmatrix}
$$

$$
\begin{bmatrix} 13 & -8 & -3 \\ -8 & 10 & -1 \\ -3 & -1 & 11 \end{bmatrix}
\begin{bmatrix} -20 \\ 5 \\ 0 \end{bmatrix}
$$

All three sets of equations can be solved separately using one of the preceding techniques. Even so, however, it is more efficient to solve all three conditions at once by using the Gauss-Jordan procedure. The coefficient matrix contains the values common to the three different circuits. However, the constant vector now becomes a constant *matrix*. Each column of the constant matrix represents a different circuit configuration and will produce the corresponding solution. The actual matrices entered into the Gauss-Jordan routine are:

$$
\begin{bmatrix} 13 & -8 & -3 \\ -8 & 10 & -1 \\ -3 & -1 & 11 \end{bmatrix}
\begin{bmatrix} 20 & 20 & -20 \\ -5 & 5 & 5 \\ 0 & 0 & 0 \end{bmatrix}
$$

The Gauss-Jordan routine returns the following values to the calling program:

$$
\begin{bmatrix} 0.19 & 0.16 & 0.07 \\ 0.16 & 0.23 & 0.06 \\ 0.07 & 0.06 & 0.11 \end{bmatrix}
\begin{bmatrix} 3 & 4.58 & -3 \\ 2 & 4.33 & -2 \\ 1 & 1.64 & -1 \end{bmatrix}
$$

The left matrix, which originally contained the coefficients, now holds the inverse of the coefficient matrix. The right matrix, which initially contained the constant vectors, now contains the corresponding solution vectors. Thus, for the three separate circuits, the answers are:

	Solution		
Circuit	I_1	I_2	I_3
1	3	2	1
2	4.58	4.33	1.64
3	-3	-2	-1

In the previous version of the Gauss-Jordan program, we copied the original coefficient matrix A into a work array B, and the constant vector Y into the vector COEF. Now we will look at a refined version of this program.

FORTRAN PROGRAM: GAUSS-JORDAN ELIMINATION WITH MULTIPLE CONSTANT VECTORS

An alternate method of using the Gauss-Jordan program is shown in Figure 4.9. The formal parameters of the Gauss-Jordan subroutine appear to be the same, but there is a subtle difference. As before, the inverse of the coefficient matrix is returned in the array that held the original coefficient matrix, but the solution vector is now a solution matrix.

Running the Program

Type up the program and execute it. You will be asked for the number of equations, as before. Next will be a question about the number of constant vectors. If this question is answered with the value of unity, then the program will behave exactly as before. However, the solution vector is now printed vertically rather than horizontally.

With this version, the number of constant vectors may be greater than one. In this example, the constants for each equation are entered in order following the corresponding coefficients. For example, if the above three electrical circuits are collectively solved with this program, then the first row of data, corresponding to the first equation, will be:

$$13 \quad -8 \quad -3 \quad 20 \quad 20 \quad -20$$

Solve all three of the above electric circuits at the same time and verify that the correct answers are obtained.

This new version has another option. The number of constant vectors may be zero. This time, only a matrix of coefficients is entered. The program will then print the inverse of the coefficient matrix. The determinant of the coefficient matrix can also be displayed if desired. It is contained in the variable DETERM.

Enter the coefficient matrix we previously had trouble with:

$$\begin{bmatrix} 1 & 1 & 1 \\ 2 & 1 & -1 \\ 3 & 1 & -3 \end{bmatrix}$$

This matrix is singular and so the determinant is zero. However, this Gauss-Jordan program is not likely to find the singularity because of round-off error. The output from this run will look like Figure 4.8. As shown in the example, you can enter zero to abort the program.

```
    Simultaneous solution by Gauss-Jordan elimination

How many equations? 3
How many constant vectors? 0

Equation   1
   1: 1
   2: 1
   3: 1

Equation   2
   1: 2
   2: 1
   3: -1

Equation   3
   1: 3
   2: 1
   3: -3

   Matrix inverse
-1.1185E+07   2.2370E+07  -1.1185E+07
 1.6777E+07  -3.3554E+07   1.6777E+07
-5.5924E+06   1.1185E+07  -5.5924E+06

How many equations? 0
```

Figure 4.8: Output from Second Version of the Gauss-Jordan Method

```
C        PROGRAM SIMQ5
C
C -- Solve simultaneous equations by
C -- Gauss-Jordan elimination with multiple constant vectors.
C -- Subroutines GAUSSJ and SWAP required.
C -- May 4, 81
C
         LOGICAL ERROR
         INTEGER SIZE, MAXR, MAXC, IN, OUT, NVEC, INDEX(8,3)
         REAL A(8,8), Y(8,3), COEF(8,3), B(8,8)
         COMMON /INOUT/ IN, OUT, MAXR, MAXC, ERROR
C
         IN = 1
         OUT = 1
         MAXR = 8
         MAXC = 8
         WRITE(OUT, 101)
10       CALL INPUT(A, Y, SIZE, NVEC)
         IF (SIZE .LT. 2) GOTO 100
         DO 30 I = 1, SIZE
            DO 20 J = 1, SIZE
               B(I,J) = A(I,J)
20       CONTINUE
```

Figure 4.9: Solution of Simultaneous Equations and
Matrix Inversion by the Gauss-Jordan Method (Multiple constant vectors may be entered.)

```
            DO 30 J = 1, NVEC
               COEF(I,J) = Y(I,J)
30       CONTINUE
         CALL GAUSSJ(B, COEF, INDEX, SIZE, MAXR, NVEC, ERROR, OUT)
         IF (.NOT. ERROR)
    *    CALL OUTPUT(A, Y, COEF, B, SIZE, NVEC)
         GOTO 10
100      STOP
101      FORMAT('1 Simultaneous solution by',
    * ' Gauss-Jordan elimination')
         END
         SUBROUTINE INPUT(A, Y, N, NVEC)
C --  Get values for N and arrays A and Y.
C
         INTEGER N, IN, OUT, I, J, M, MAXR, NVEC
         REAL A(8,8), Y(8,3)
         COMMON /INOUT/ IN, OUT, MAXR
C
5        WRITE(OUT, 105)
         READ(IN, 106) N
         M = N
         IF (N .GT. MAXR) GOTO 5
         IF (N .LT. 2) RETURN
7        WRITE(OUT, 107)
         READ(IN, 106) NVEC
         IF ((NVEC .GT. 3) .OR. (NVEC.LT.0)) GOTO 7
         DO 30 I = 1, N
            WRITE(OUT, 101) I
            DO 10 J = 1, N
               WRITE(OUT, 102) J
               READ(IN, 103) A(I,J)
10          CONTINUE
            IF (NVEC .EQ. 0) GOTO 30
            DO 20 J = 1, NVEC
               WRITE(OUT, 104)
               READ(IN, 103) Y(I,J)
20          CONTINUE
30       CONTINUE
         RETURN
101      FORMAT(' Equation ', I3/)
102      FORMAT('+',I4, ': ')
103      FORMAT(F10.0)
104      FORMAT('+ C: ')
105      FORMAT(/' How many equations? ')
106      FORMAT(I2)
107      FORMAT(' How many constant vectors? ')
         END
         SUBROUTINE OUTPUT(A, Y, COEF, INVER, N, NVEC)
C
C --  Print the answers.
C
         LOGICAL ERROR
         INTEGER N, IN, OUT, I, J, MAXR, MAXC, NVEC
```

Figure 4.9: Solution of Simultaneous Equations and
Matrix Inversion by the Gauss-Jordan Method (Multiple constant vectors may be entered.) (cont.)

```
         REAL A(8,8), Y(8,3), COEF(8,3), INVER(8,8)
         COMMON /INOUT/ IN, OUT, MAXR, MAXC, ERROR
C
         IF (NVEC .EQ. 0) GOTO 30
C -- Print matrix and solution vector.
         DO 10 I = 1, N
            WRITE(OUT, 101)
     *      (A(I,J), J = 1, N), (Y(I,J), J = 1, NVEC)
10       CONTINUE
         IF (ERROR) RETURN
         WRITE(OUT, 102)
         DO 20 I = 1, N
            WRITE(OUT, 101) (COEF(I,J), J = 1, NVEC)
20       CONTINUE
         RETURN
C -- Print inverse.
30       WRITE(OUT, 103)
         DO 40 I = 1, N
            WRITE(OUT, 101)
     *      (INVER(I,J), J = 1, N)
40       CONTINUE
         RETURN
101      FORMAT(1P6E12.4)
102      FORMAT(/ '    Solution')
103      FORMAT(/ '    Matrix inverse')
         END
```

Figure 4.9: Solution of Simultaneous Equations and Matrix Inversion by the Gauss-Jordan Method (Multiple constant vectors may be entered.) (cont.)

In the following section we will discuss the Hilbert matrix as an example of ill conditioning. To explore such matrices we will write a program that uses our first version of the Gauss-Jordan method. We will then use the program to solve a series of progressively more problematic Hilbert matrices.

ILL-CONDITIONED EQUATIONS

A singular matrix has a determinant of zero. The corresponding set of equations has either no solutions or many solutions. An *ill-conditioned matrix*, by comparison, is one that is *nearly* singular. It produces incorrect or inaccurate answers. A small change in the input data can cause great changes in the answer. A two-dimensional analogue of ill conditioning occurs with two nearly parallel lines. The point of intersection is the desired solution, but the closer the two lines come to being parallel, the more difficult it is to determine their actual point of intersection.

Various tests for ill conditioning have been suggested. One of these is to compare the values of the inverted matrix to those of the original matrix. If there are differences of several orders of magnitude, then it is likely that ill conditioning is present. Another test is to take the inverse of the inverse of the coefficient matrix and compare the result to the original matrix. They should, of course, be the same. This technique will test the inversion algorithm and the computer arithmetic at the same time. Yet another test is to compare the magnitudes of values along the major diagonal. They should not be too far apart.

The Hilbert matrix is an example of ill conditioning. This symmetric matrix begins with unity in the upper-left corner. The remaining values get smaller and smaller as we go down a column or across a row, according to the pattern:

$$
\begin{bmatrix}
1 & \frac{1}{2} & \frac{1}{3} & \frac{1}{4} & \cdots\cdots\cdots & 1/n \\
\frac{1}{2} & \frac{1}{3} & \frac{1}{4} & \frac{1}{5} & \cdots & 1/(n+1) \\
\frac{1}{3} & \frac{1}{4} & \frac{1}{5} & \frac{1}{6} & \cdots & 1/(n+2) \\
\frac{1}{4} & \frac{1}{5} & \frac{1}{6} & \frac{1}{7} & \cdots & 1/(n+3) \\
\cdots & & \cdots & & & \cdots \\
1/n & \cdots\cdots\cdots\cdots\cdots & & & & 1/(2n-1)
\end{bmatrix}
$$

The Hilbert matrix can be used to produce a set of ill-conditioned equations. Consider, for example:

$$x_1 + \frac{x_2}{2} + \frac{x_3}{3} + \frac{x_4}{4} + \frac{x_5}{5} = 1 + \tfrac{1}{2} + \tfrac{1}{3} + \tfrac{1}{4} + \tfrac{1}{5}$$

$$\frac{x_1}{2} + \frac{x_2}{3} + \frac{x_3}{4} + \frac{x_4}{5} + \frac{x_5}{6} = \tfrac{1}{2} + \tfrac{1}{3} + \tfrac{1}{4} + \tfrac{1}{5} + \tfrac{1}{6}$$

$$\frac{x_1}{3} + \frac{x_2}{4} + \frac{x_3}{5} + \frac{x_4}{6} + \frac{x_5}{7} = \tfrac{1}{3} + \tfrac{1}{4} + \tfrac{1}{5} + \tfrac{1}{6} + \tfrac{1}{7}$$

$$\frac{x_1}{4} + \frac{x_2}{5} + \frac{x_3}{6} + \frac{x_4}{7} + \frac{x_5}{8} = \tfrac{1}{4} + \tfrac{1}{5} + \tfrac{1}{6} + \tfrac{1}{7} + \tfrac{1}{8}$$

$$\frac{x_1}{5} + \frac{x_2}{6} + \frac{x_3}{7} + \frac{x_4}{8} + \frac{x_5}{9} = \tfrac{1}{5} + \tfrac{1}{6} + \tfrac{1}{7} + \tfrac{1}{8} + \tfrac{1}{9}$$

These equations meet all our criteria for ill-conditioning. First, the inverse of the inverse will not equal the original matrix, because the fractions 1/3, 1/6 and 1/7 cannot be represented exactly. Consequently, there will be roundoff errors at the very beginning of the problem. Second, the inverse of the coefficient matrix is exactly:

$$\begin{bmatrix} 25 & -300 & 1050 & -1400 & 630 \\ -300 & 4800 & -18900 & 26880 & -12600 \\ 1050 & -18900 & 79380 & -117600 & 56700 \\ -1400 & 26880 & -117600 & 179200 & -88200 \\ 630 & -12600 & 56700 & -88200 & 44100 \end{bmatrix}$$

Some of the elements of the inverse are orders of magnitude larger than elements of the original matrix. Furthermore, the determinant of the coefficient matrix is nearly zero.

Since each element of the constant vector is the sum of the matrix elements in the corresponding row, the exact solution is:

$$\begin{bmatrix} 1 & 1 & 1 & 1 & 1 \end{bmatrix}$$

But, because of the ill conditioning, the calculated solution might be something like this:

$$\begin{bmatrix} 1.0001 & 0.99816 & 1.00778 & 0.9884 & 1.0056 \end{bmatrix}$$

Now let us look at a program that uses Hilbert matrices to study the effect of ill conditioning.

FORTRAN PROGRAM: SOLVING HILBERT MATRICES

The program shown in Figure 4.10 will generate a set of Hilbert matrices and constant vectors corresponding to a solution of:

$$\begin{bmatrix} 1 & 1 & 1 & \ldots & 1 \end{bmatrix}$$

Running the Program

Type up the program and execute it. It will run automatically. The program begins with the two equations:

$$\begin{bmatrix} 1 & \frac{1}{2} \\ \frac{1}{2} & \frac{1}{3} \end{bmatrix} \qquad \begin{bmatrix} \frac{3}{2} \\ \frac{5}{6} \end{bmatrix}$$

which are solved by calling the first version of the Gauss-Jordan subroutine. A matrix this small is not ill conditioned, and so no problem is apparent. The solution vector is [1 1]. The program then continues with three, four, five, six, and seven equations.

If your FORTRAN floating-point operations are performed with the usual 32-bit precision, you will start seeing roundoff errors with four equations. Six equations will give accuracy of only two or three significant figures, and the results for seven equations will be meaningless.

Alternately, if you have double-precision, 64-bit floating-point operations, then the situation is very different. Now, the solution to seventeen simultaneous equations will be given to an accuracy of seven significant figures or better. This program can thus be used to test the significance of your floating-point package.

```
C       PROGRAM SIMQ3
C
C -- Solution of simultaneous Hilbert matrices
C -- by Gauss-Jordan elimination.
C -- Subroutines GAUSSJ and SWAP needed.
C -- May 4, 81
C
        LOGICAL ERROR
        INTEGER SIZE, MAXR, MAXC, IN, OUT, INDEX(8,3), NVEC
        REAL A(8,8), Y(8), COEF(8), B(8,8)
        COMMON /INOUT/ IN, OUT, MAXR, MAXC
        DATA NVEC/1/
C
        IN = 1
        OUT = 1
        MAXR = 8
        MAXC = 8
        WRITE(OUT, 101)
        A(1,1) = 1.0
        SIZE = 2
10      CALL INPUT(A, Y, SIZE)
        DO 30 I = 1, SIZE
          DO 20 J = 1, SIZE
            B(I,J) = A(I,J)
20        CONTINUE
          COEF(I) = Y(I)
30      CONTINUE
        CALL GAUSSJ(B, COEF, INDEX, SIZE, MAXR, NVEC, ERROR, OUT)
        IF (.NOT. ERROR) CALL OUTPUT(A, Y, COEF, SIZE)
        SIZE = SIZE + 1
        IF (SIZE .LE. MAXR) GOTO 10
        WRITE(OUT, 102)
        STOP
101     FORMAT('1 Solution of Hilbert matrices by',
      * ' Gauss-Jordan elimination')
102     FORMAT(/)
        END
        SUBROUTINE INPUT(A, Y, N)
C
C --   Get values for N and arrays A and Y.
C
        INTEGER N, IN, OUT, I, J, M
        REAL A(8,8), Y(8)
        COMMON /INOUT/ IN, OUT, MAXR
C
```

Figure 4.10: Solution of a Set of Ill-Conditioned Equations

```
        DO 10 I = 1, N
          A(N,I) = 1.0 / (N + I - 1)
          A(I,N) = A(N,I)
10      CONTINUE
        A(N,N) = 1.0 / (2*N - 1)
        DO 20 I = 1, N
          Y(I) = 0.0
          DO 20 J = 1, N
            Y(I) = Y(I) + A(I,J)
20      CONTINUE
        RETURN
        END
        SUBROUTINE OUTPUT(A, Y, COEF, N)
C
C -- Print the answers.
C
        INTEGER N, IN, OUT, I, J
        REAL A(8,8), Y(8), COEF(8)
        COMMON /INOUT/ IN, OUT
C
        IF (N .GT. 4) GOTO 20
        WRITE(OUT, 103)
        DO 10 I = 1, N
          WRITE(OUT, 101) (A(I,J), J = 1, N), Y(I)
10      CONTINUE
20      WRITE(OUT, 102)
        WRITE(OUT, 101) (COEF(I), I = 1, N)
        RETURN
101     FORMAT(8F9.5)
102     FORMAT(/ '    Solution')
103     FORMAT(/)
        END
```

Figure 4.10: Solution of a Set of Ill-Conditioned Equations (cont.)

In the next section we will see another variation of the Gauss-Jordan subroutine. We will consider a set of equations in which the number of equations is greater than the number of variables per equation, and we will present a program to compute the "best-fit" solution.

A SIMULTANEOUS BEST FIT

The previous programs in this chapter produce an *exact* fit to a set of linear equations. In each case, the number of unknowns equals the number of independent equations. If, however, there are more unknowns than there are equations, then no unique solution is possible. This situation corresponds to a coefficient matrix having more columns than rows.

There is another case we might consider. Suppose that we want to determine the value of m unknowns by an experimental procedure. If m independent measurements are obtained, then we can find an exact solution. On the other hand, suppose that the number of independent measurements, n, is greater than the number of unknowns, m. Then it is possible to calculate a *best-fit* value for the m unknowns.

As a two-dimensional analogue, consider the three equations:

$$
\begin{aligned}
x + y &= 3 \\
x &= 1 \\
y &= 1
\end{aligned}
$$

These three lines do not intersect at the same point. Rather, each pair of lines defines a different point on the x-y plane. These points describe a right triangle that has corners at the positions (1, 1), (2, 1), and (1, 2), as shown in Figure 4.11. The best choice for the "intersection" of these three lines is the location of the centroid of the triangle. Since this point is located at one-third of the distance from the base to the apex, then the best-fit solution to these three lines is:

$$x = 1.3333, \quad y = 1.3333$$

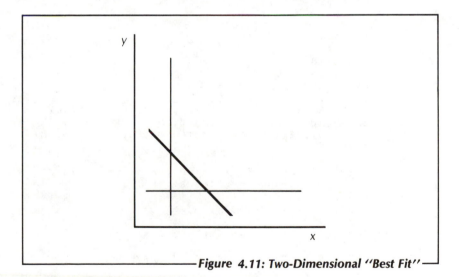

Figure 4.11: Two-Dimensional "Best Fit"

Now that we have illustrated a best-fit solution, we will look at the program that will find the solution.

FORTRAN PROGRAM: THE BEST-FIT SOLUTION

Make a copy of the program shown in Figure 4.6 and alter it to look like Figure 4.12. Subroutine SQUARE, developed in the previous chapter, is included. It is used to convert the rectangular array of coefficients and the constant vector into the square array and vector needed by subroutine

GAUSSJ. Subroutines GAUSSJ and SWAP are not shown, but they must be included also.

The technique used in this program is essentially the same as that used for *least-squares curve fitting*. Since this topic is covered more fully in the next chapter, it will not be further discussed here.

```
C       PROGRAM SIMQ5
C
C -- FORTRAN program to solve simultaneous equations
C -- by Gauss-Jordan elimination.
C -- There may be more equations than unknowns.
C -- Subroutines SQUARE, GAUSSJ and SWAP are also needed.
C -- May 4, 81
C
        LOGICAL ERROR
        INTEGER MAXR, MAXC, IN, OUT, N, M, I, J, INDEX(8,3), NVEC
        REAL A(8,8), Y(8), COEF(8), B(8,8), YY(8)
        COMMON /INOUT/ IN, OUT, MAXR, MAXC, ERROR
        DATA NVEC/1/
C
        IN = 1
        OUT = 1
        MAXR = 8
        MAXC = 8
        WRITE(OUT, 101)
10      CALL INPUT(A, Y, N, M)
        IF (M .LT. 2) GOTO 100
        CALL SQUARE(A, Y, B, COEF, N, M, MAXR, MAXC)
        CALL GAUSSJ(B, COEF, INDEX, M, MAXR, NVEC, ERROR, OUT)
        IF (.NOT. ERROR) CALL OUTPUT(A, Y, COEF, N, M)
        GOTO 10
100     STOP
101     FORMAT('1 Best fit to simultaneous equations',
       * ' by Gauss-Jordan elimination')
        END
        SUBROUTINE INPUT(A, Y, N, M)
C
C --  Get values for N and arrays A and Y.
C
        INTEGER N, M, IN, OUT, I, J, MAXR
        REAL A(8,8), Y(8)
        COMMON /INOUT/ IN, OUT, MAXR, MAXC
C
5       WRITE(OUT, 107)
        READ(IN, 106) M
        IF (M .GT. MAXC) GOTO 5
        IF (M .LT. 2) RETURN
7       WRITE(OUT, 105)
        READ(IN, 106) N
        IF (N .LT. M) GOTO 7
        DO 20 I = 1, N
```

Figure 4.12: The Best Fit to a Set of Linear Equations

```
                WRITE(OUT, 101) I
                DO 10 J = 1, M
                   WRITE(OUT, 102) J
                   READ(IN, 103) A(I,J)
      10        CONTINUE
                WRITE(OUT, 104)
                READ(IN, 103) Y(I)
      20     CONTINUE
           RETURN
     101    FORMAT(' Equation ', I3/)
     102    FORMAT('+',I4, ': ')
     103    FORMAT(F10.0)
     104    FORMAT('+ C: ')
     105    FORMAT(' How many equations? ')
     106    FORMAT(I2)
     107    FORMAT(/' How many unknowns? ')
           END
           SUBROUTINE OUTPUT(A, Y, COEF, N, M)
   C
   C -- Print the answers.
   C
           LOGICAL ERROR
           INTEGER N, M, IN, OUT, I, J, MAXR, MAXC
           REAL A(8,8), Y(8), COEF(8)
           COMMON /INOUT/ IN, OUT, MAXR, MAXC, ERROR
   C
           DO 10 I = 1, N
              WRITE(OUT, 101) (A(I,J), J = 1, M), Y(I)
      10     CONTINUE
           WRITE(OUT, 102)
           IF (ERROR) RETURN
           WRITE(OUT, 101) (COEF(I), I = 1, M)
           RETURN
     101    FORMAT(1P6E12.4)
     102    FORMAT(/ '    Solution')
           END
```

Figure 4.12: The Best Fit to a Set of Linear Equations (cont.)

Running the Best-Fit Program

Compile the program and execute it. First, the number of unknowns is requested, then the number of equations. If these have the same value, then a regular, simultaneous solution is performed. However, if there are more equations than unknowns, then the best fit is returned. The three equations that we discussed earlier in this section have only two unknowns. The matrix and vector are:

$$\begin{bmatrix} 1 & 1 \\ 1 & 0 \\ 0 & 1 \end{bmatrix} \quad \begin{bmatrix} 3 \\ 1 \\ 1 \end{bmatrix}$$

Verify that the result is:

$$x = 1.3333, \quad y = 1.3333$$

As another example, consider the electric circuit from the beginning of this chapter. We originally derived three loop-current equations that were solved simultaneously. Suppose, however, that the voltages of the sources were determined experimentally. If the value of the left source was found to be 19 volts, and the value of the right source was measured as -5.1 volts, then the three loop equations would be:

$$\begin{bmatrix} 13 & -8 & -3 \\ -8 & 10 & -1 \\ -3 & -1 & 11 \end{bmatrix} \quad \begin{bmatrix} 19 \\ -5.1 \\ 0 \end{bmatrix}$$

In addition, suppose that the voltage across the horizontal, one-ohm resistor was measured to be 1.1 volts. The current flowing through this resistor is loop current 2 minus loop current 3. Consequently, we can now write an additional independent equation, corresponding to an additional row in our matrix. The four equations are:

$$\begin{bmatrix} 13 & -8 & -3 \\ -8 & 10 & -1 \\ -3 & -1 & 11 \\ 0 & 1 & -1 \end{bmatrix} \quad \begin{bmatrix} 19 \\ -5.1 \\ 0 \\ 1.1 \end{bmatrix}$$

Enter these four equations, with their three unknowns, into the new program. Compare the resulting, best-fit solution:

$$I_1 = 2.8 \text{ amps}, \quad I_2 = 1.8 \text{ amps}, \quad I_3 = 0.94 \text{ amps}$$

to the original solution. This program can be aborted by entering zero for the number of equations.

Next we will devise a method—and a FORTRAN program—for solving simultaneous equations that have complex coefficients (i.e., factors containing imaginary parts). To illustrate this problem we will study a second, more complicated electrical example.

EQUATIONS WITH COMPLEX COEFFICIENTS

Simultaneous equations with complex coefficients occur in the analysis of electrical circuits. Complex numbers are not always incorporated into FORTRAN compilers. Nevertheless, it is rather easy to solve n complex simultaneous equations by converting them into $2n$ equations with real coefficients. The resulting equations can then be solved by one of the methods developed previously in this chapter.

Example: An Alternating-Current Electrical Circuit

Consider the electrical circuit shown in Figure 4.13. This circuit is more complicated than Figure 4.1, since it contains an AC power source, an inductor, and a capacitor. The impedance for a resistor is simply the resistance R. However, the impedance for inductors and capacitors is a function of the frequency. The impedance function for the inductor is $j\omega L$, where j is the imaginary operator equal to the square root of -1, ω is the frequency of the AC source in radians per second, and L is the self-inductance in henries. The impedance function for the capacitor is $-j/\omega C$, where C is the capacitance in farads.

For Figure 4.13, the impedance functions are shown next to the corresponding elements. The AC power source is 10 volts (RMS) with a frequency of ω. The phase angle is arbitrarily chosen to be zero. The inductor has an impedance of $j5$ ohms and the capacitor has an impedance of $-j4$ ohms.

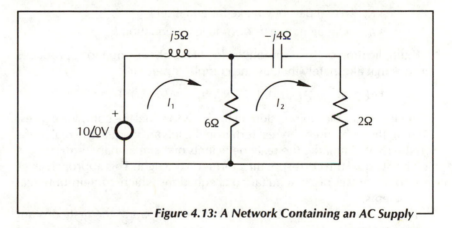

Figure 4.13: A Network Containing an AC Supply

We can find the branch currents for this circuit by using the two loop currents and the Kirchhoff voltage law. A clockwise summing around each loop gives:

$$(6 + j5)I_1 - 6I_2 - 10 \;=\; 0 \quad \text{(left loop)}$$

$$-6I_1 + (8 - j4)I_2 \;=\; 0 \quad \text{(right loop)}$$

The corresponding linear equations are:

$$\begin{bmatrix} (6 + j5) & (-6 + j0) \\ (-6 + j0) & (8 - j4) \end{bmatrix} \quad \begin{bmatrix} (10 + j0) \\ (0 + j0) \end{bmatrix}$$

These two equations cannot be directly solved with the programs given previously in this chapter because they contain complex coefficients.

Let us consider a general statement of the two loop equations. The electrical current and the impedance can both be expressed as complex numbers. Consequently, we can write:

$$(AR_{11} + jAI_{11})\ (IR_1 + jII_1) + (AR_{21} + jAI_{21})\ (IR_2 + jII_2)$$
$$= (VR_1 + jVI_1)$$

$$(AR_{21} + jAI_{21})\ (IR_1 + jII_1) + (AR_{22} + jAI_{22})\ (IR_2 + jII_2)$$
$$= (VR_2 + jVI_2)$$

where the following symbols are used:

AR_{kl} = real part of coefficient (impedance) k,l
AI_{kl} = imaginary part of coefficient k,l
IR_l = real part of current l
II_l = imaginary part of current l
VR_k = real part of voltage for equation k
VI_k = imaginary part of voltage for equation k

Multiplication of the terms on the left of the above equations produces groups that alternately include the complex operator j.

$$(AR_{11}\ IR_1 - AI_{11}\ II_1) + j(AR_{11}\ II_1 + AI_{11}\ IR_1) + \ldots$$

But if the complex expression on the left is to equal the complex expression on the right, then the real terms on the left must equal the real terms on the right. Similarly, the real coefficients of the imaginary terms on the left must equal the corresponding terms on the right. This approach gives rise to a new set of $2n$ simultaneous equations which contain only real coefficients.

The first new equation is set equal to the real part of the first constant (voltage) term:

$$AR_{11} - AI_{11} + AR_{12} - AI_{12} = VR_1$$

Notice that the complex conjugates of the original coefficients appear in the new first equation. That is, the original coefficients appear in order, but with alternating signs.

The second new equation is set equal to the imaginary part of the first constant (voltage) term:

$$AI_{11} + AR_{11} + AI_{12} + AR_{12} = VI_1$$

This equation also contains all of the coefficients for the first original equation. But in this case, the real and imaginary parts are interchanged. Furthermore, the original signs are utilized.

The complete new equations can be summarized as:

$$\begin{bmatrix} AR_{11} & -AI_{11} & AR_{12} & -AI_{12} \\ AI_{11} & AR_{11} & AI_{12} & AR_{12} \\ AR_{21} & -AI_{21} & AR_{22} & -AI_{22} \\ AI_{21} & AR_{21} & AI_{22} & AR_{22} \end{bmatrix} \quad \begin{bmatrix} VR_1 \\ VI_1 \\ VR_2 \\ VI_2 \end{bmatrix}$$

Substituting the values from Figure 4.13 gives:

$$\begin{bmatrix} 6 & -5 & -6 & 0 \\ 5 & 6 & 0 & -6 \\ -6 & 0 & 8 & 4 \\ 0 & -6 & -4 & 8 \end{bmatrix} \quad \begin{bmatrix} 10 \\ 0 \\ 0 \\ 0 \end{bmatrix}$$

Notice that each original coefficient appears twice in the new matrix. The solution vector for the new set of equations can readily be found by the methods of this chapter. The solution is:

$$[1.5 \quad -2.0 \quad 1.5 \quad -0.75]$$

which corresponds to the loop currents:

$$I_1 = 1.5 - j2 \text{ amps} = 2.5 \ \underline{/-53°}$$
$$I_2 = 1.5 - j0.75 \text{ amps} = 1.67 \ \underline{/-27°}$$

These results can be readily verified by calculating the voltages across each circuit element. For example, if the lower node is chosen to be zero volts, then the voltage of the upper node is equal to the voltage across the 6-ohm resistor:

$$V = 6(I_1 - I_2) = 6(-j1.25) = -j7.5 \text{ volts}$$

Similarly, the voltage across the inductor is:

$$V = j5(I_1) = 10 + j7.5 \text{ volts}$$

A sum of the voltages around the left loop then gives:

$$(-j7.5) + (10 + j7.5) - (10) = 0$$

A similar check can be made on the right loop.

Let us now look at a program that will handle these complex coefficients.

FORTRAN PROGRAM: SIMULTANEOUS EQUATIONS WITH COMPLEX COEFFICIENTS

The program given in Figure 4.14 simplifies the solution of simultaneous equations with complex coefficients. Each coefficient of the original n equations is entered only once. Then the program converts the data into a $2n$-by-$2n$ matrix and a constant vector of length $2n$. Up to four complex equations can be solved simultaneously. This number can be increased by changing the values of MAXR and MAXC. (They must be no bigger than half the dimensioned size of the arrays A, Y, and COEF.) The value of 8 in the declaration statements must also be increased to twice the value of MAXR.

Any of the methods previously developed in this chapter can find the solution. We have selected the Gauss-Jordan technique. Consequently, it might be best to begin with a copy of the Gauss-Jordan program given in Figure 4.6.

```
C       PROGRAM SIMQ11
C
C -- Solve simultaneous equations with complex coefficients
C -- by Gauss-Jordan elimination.
C -- Subroutines GAUSSJ and SWAP required.
C -- May 4, 81
C
        LOGICAL ERROR
        INTEGER N, M, MAXR, MAXC, IN, OUT, I, J, INDEX(8,3), NVEC
        REAL A(8,8), Y(8), COEF(8), B(8,8)
        COMMON /INOUT/ IN, OUT, MAXR, MAXC, ERROR
        DATA NVEC/1/
C
        IN = 1
        OUT = 1
        MAXR = 4
        MAXC = 4
        WRITE(OUT, 101)
10      CALL INPUT(A, Y, N, M)
        IF (N .LT. 2) GOTO 100
        DO 30 I = 1, N
          DO 20 J = 1, N
            B(I,J) = A(I,J)
20        CONTINUE
          COEF(I) = Y(I)
30      CONTINUE
        CALL GAUSSJ(B, COEF, INDEX, N, MAXR*2, NVEC, ERROR, OUT)
        IF (.NOT. ERROR) CALL OUTPUT(A, Y, COEF, N)
        GOTO 10
100     STOP
101     FORMAT('1 Simultaneous solution',
      * ' with complex coefficients')
        END
        SUBROUTINE INPUT(A, Y, N, M)
C -- Get values for N and arrays A and Y.
C
        INTEGER N, IN, OUT, I, J, M, MAXR, K, L
        REAL A(8,8), Y(8), DREAL(4,4), DIMAG(4,4), V(4,2)
        COMMON /INOUT/ IN, OUT, MAXR
C
5       WRITE(OUT, 105)
        READ(IN, 106) N
```

Figure 4.14: Simultaneous Solution of Equations with Complex Coefficients

```
          M = N
          IF (N .GT. MAXR) GOTO 5
          IF (N .LT. 2) RETURN
          DO 20 I = 1, N
            WRITE(OUT, 101) I
            K = 0
            L = 2 * I - 1
            DO 10 J = 1, N
              K = K + 1
              WRITE(OUT, 102) J
              READ(IN, 103) DREAL(I,J)
              A(L,K) = DREAL(I,J)
              A(L+1, K+1) = DREAL(I,J)
              K = K + 1
              WRITE(OUT, 107) J
              READ(IN, 103) DIMAG(I,J)
              A(L,K) = -DIMAG(I,J)
              A(L+1, K-1) = DIMAG(I,J)
10          CONTINUE
            WRITE(OUT, 104)
            READ(IN, 103) V(I,1)
            Y(L) = V(I,1)
            WRITE(OUT, 108)
            READ(IN, 103) V(I,2)
            Y(L+1) = V(I,2)
20        CONTINUE
C -- Print original matrix.
          DO 30 I = 1, N
            WRITE(OUT, 109) (DREAL(I,J), DIMAG(I,J), J = 1, N),
          *   V(I,1), V(I,2)
30        CONTINUE
          WRITE(OUT, 110)
          N = 2 * N
          M = N
          RETURN
101       FORMAT(' Equation ', I3/)
102       FORMAT('+Real ', I3, ': ')
103       FORMAT(F10.0)
104       FORMAT('+Real constant ')
105       FORMAT(/' How many equations? ')
106       FORMAT(I2)
107       FORMAT('+Imag ', I3, ': ')
108       FORMAT('+Imag constant ')
109       FORMAT(1P6E12.4)
110       FORMAT(/)
          END
          SUBROUTINE OUTPUT(A, Y, COEF, N)
C
C -- Print the answers.
C
          LOGICAL ERROR
          INTEGER N, IN, OUT, I, J, MAXR, MAXC, N2
```

Figure 4.14: Simultaneous Solution of Equations with Complex Coefficients (cont.)

```
          REAL A(8,8), Y(8), COEF(8), RE, IM, MAG, ANG, P180
          COMMON /INOUT/ IN, OUT, MAXR, MAXC, ERROR
          DATA P180/57.29578/
   C
          DO 10 I = 1, N
            WRITE(OUT, 101) (A(I,J), J = 1, N), Y(I)
   10     CONTINUE
          IF (ERROR) RETURN
          N2 = N / 2
          WRITE(OUT, 102)
          DO 20 I = 1, N2
            J = 2 * I - 1
            RE = COEF(J)
            IM = COEF(J + 1)
            MAG = SQRT(RE*RE + IM*IM)
            ANG = ATAN2(IM, RE) * P180
            WRITE(OUT, 103) RE, IM, MAG, ANG
   20     CONTINUE
          RETURN
   101    FORMAT(1P6E12.4)
   102    FORMAT(/ '     Real          Imaginary',
         * '     Magnitude      Angle')
   103    FORMAT(1P4E12.4)
          END
```

Figure 4.14: Simultaneous Solution of Equations with Complex Coefficients (cont.)

Running the Program

Type up the program and use it to solve the circuit shown in Figure 4.13. Enter the value of 2 for the number of complex equations. Then enter the coefficients for each equation in turn. Give the real coefficient first, then the imaginary part next. The constant vector is entered in the same way; real part first, then imaginary part. The data for this problem are entered as:

$$6, 5, -6, \quad 0, 10, 0 \text{ (equation 1)}$$
$$-6, 0, \quad 8, -4, \quad 0, 0 \text{ (equation 2)}$$

The original input data are printed out, and the new 2n-by-2n matrix is given. Then, the solution is given in both rectangular and polar forms.

The polar magnitude (MAG) is found from the square root of the sum of the squares of the rectangular components. The phasor angle (ANG) is calculated by the FORTRAN function ATAN2. Since this function takes two arguments, the resulting angle is placed into the proper quadrant. Multiplication by $180/\pi$ converts the result from radians to degrees.

We have now seen several methods and variations that are adequate for solving small matrices (i.e., small sets of simultaneous equations). The last topic of this chapter will be an iterative method for solving large matrices.

THE GAUSS-SEIDEL ITERATIVE METHOD

The Gauss elimination and Gauss-Jordan methods we considered previously are not suitable for solving very large matrices. More and more multiplication and subtraction operations are performed as the number of equations increases. The resulting roundoff error can produce a meaningless solution.

The Gauss-Seidel method finds the solution to a set of equations by an iterative technique. An initial approximation is repeatedly refined until the result is acceptably close to the solution. Since each approximation depends only on the previous approximation, roundoff error does not accumulate. An added feature is that the equations do not have to be linear.

Consider the three loop-current equations derived from Figure 4.1:

$$13I_1 - 8I_2 - 3I_3 = 20$$
$$-8I_1 + 10I_2 - I_3 = -5$$
$$-3I_1 - I_2 + 11I_3 = 0$$

We can solve the first equation for I_1:

$$I_1 = \frac{20 + 8I_2 + 3I_3}{13}$$

in terms of the other two unknowns. Then if we chose first approximations of zero for I_2 and I_3, we obtain a value of 1.54 for I_1. The second equation is then solved for the second variable:

$$I_2 = \frac{8I_1 + I_3 - 5}{10}$$

Substituting the current values of 1.54 for I_1 and zero for I_3 produces a value of 0.73 for I_2. The third equation is similarly solved for the third variable:

$$I_3 = \frac{3I_1 + I_2}{11}$$

The current values of 1.54 for I_1 and 0.73 for I_2 give a value of 0.486 for I_3. The process is now repeated. The values of 0.73 for I_2 and 0.486 for I_3 are used to obtain a better value for I_1. After about 20 complete iterations, the values are correct to three significant figures. The following table gives some of the values in the sequence:

I_1	I_2	I_3
0	0	0
1.54	0.73	0.486
2.1	1.23	0.685
2.45	1.53	0.808
2.67	1.71	0.883

There are several potential problems with the Gauss-Seidel method. First of all, the process might not converge. That is, successive values may drift further and further from the correct solution. Consider, for example, the previous three equations. If they are written in reverse order, then we will derive the expressions:

$$I_1 = (11 I_3 - I_2) / 3$$
$$I_2 = (13 I_1 - 3 I_3 - 20) / 13$$
$$I_3 = 10 I_2 - 8 I_1$$

First approximations of zero are chosen, as before. However, the subsequent values are clearly diverging:

I_1	I_2	I_3
0	0	0
0	−1.5	−15
−57	−54	−92
−318	−299	−440

The problem in this second case is that the largest values are not located on the major diagonal. The solution is to interchange rows to bring the largest element into the pivot position.

Finally, let us investigate a FORTRAN implementation of the Gauss-Seidel method. We will make note of a couple of interesting features in this program:

- the differences between *relative* and *absolute criteria* in IF statements.
- the meaning and use of *point relaxation*.

FORTRAN PROGRAM: THE GAUSS-SEIDEL METHOD

Surprisingly, the choice of a first approximation is not too important. An additional matter to be considered, however, is the criterion for convergence. The program shown in Figure 4.15 can be used to explore the Gauss-Seidel method for the solution of linear simultaneous equations.

Most of the program can be derived from the Gaussian elimination method given in Figure 4.4.

The pivot-interchange routine used in the Gauss elimination program is incorporated here for the same purpose. Rows are interchanged to place the largest element of each column on the major diagonal.

An absolute, rather than a relative, criterion is used to determine convergence:

IF (ABS(NEXTC — COEF(J)) .LT. TOL) GOTO 95

Normally, it is better to chose a relative criterion:

IF (ABS(1 — COEF(J) / NEXTC) .LT. TOL) GOTO 95

In this case, however, we must insure that NEXTC will never be zero. One way to do this would be to use the form:

IF (ABS(NEXTC — COEF(J)) .LT. ABS(TOL＊NEXTC)) GOTO 95

Sometimes the successive approximations jump about wildly. One feature of the program given in Figure 4.15 will reduce this tendency. If two successive approximations differ in sign, then the step size is cut in half.

Another feature of the Gauss-Seidel program shown in Figure 4.15 is known as point relaxation. With this technique, each selected value is a function of the previous iteration, the calculated value, and a relaxation factor, lambda. If COEF(J) is the previous value and NEXTC is the calculated value, then the actual next value becomes:

COEF(J) = LAMBDA ＊ NEXTC + (1 — LAMBDA)＊COEF(J)

The value of LAMBDA can range from 0 to 2.

```
C       PROGRAM SIMQ12
C
C -- Solve simultaneous equations by Gauss-Seidel method.
C -- Apr 30, 81
C
        LOGICAL ERROR
        INTEGER LENGTH, MAXR, MAXC, IN, OUT
        REAL A(8,8), Y(8), COEF(8)
        COMMON /INOUT/ IN, OUT, MAXR, MAXC
C
        IN = 1
        OUT = 1
        MAXR = 8
        MAXC = 8
        WRITE(OUT, 101)
```

Figure 4.15: Solution of Linear Equations by the Gauss-Seidel Method

```
10        CALL INPUT(A, Y, LENGTH)
          IF (LENGTH .LT. 2) STOP`
          CALL SEID(A, Y, COEF, LENGTH, ERROR)
          IF (.NOT. ERROR) CALL OUTPUT(A, Y, COEF, LENGTH)
          GOTO 10
101       FORMAT('1 Simultaneous solution by Gauss-Seidel')
          END
          SUBROUTINE INPUT(A, Y, N)
C --   Get values for N and arrays A and Y.
C
          INTEGER N, IN, OUT, I, J, M
          REAL A(8,8), Y(8)
          COMMON /INOUT/ IN, OUT, MAXR
C
5         WRITE(OUT, 105)
          READ(IN, 106) N
          M = N
          IF (N .GT. MAXR) GOTO 5
          IF (N .EQ. 0) RETURN
          IF (N .LT. 0) GOTO 30
          DO 20 I = 1, N
            WRITE(OUT, 101) I
            DO 10 J = 1, N
              WRITE(OUT, 102) J
              READ(IN, 103) A(I,J)
10          CONTINUE
            WRITE(OUT, 104)
            READ(IN, 103) Y(I)
20        CONTINUE
          RETURN
C -- Use matrix from previous problem.
30        N = -N
          RETURN
101       FORMAT(' Equation ', I3/)
102       FORMAT('+',I4, ': ')
103       FORMAT(F10.0)
104       FORMAT('+ C: ')
105       FORMAT(/' How many equations? ')
106       FORMAT(I2)
          END
          SUBROUTINE SEID(A, Y, COEF, NCOL, ERROR)
C
C -- Solve simultaneous equations by Gauss-Seidel method.
C -- Apr 28, 81
C
          LOGICAL ERROR, DONE
          INTEGER NCOL, I, J, IN, OUT, N, N1, K, L, MAX
          REAL A(8,8), Y(8), COEF(8), TOL, HOLD, LAMBDA
          REAL B(8,8), W(8), BIG, AB, SUM, T, NEXTC
          COMMON /INOUT/ IN, OUT
          DATA TOL/ 1.0E-4/, MAX/100/
C
```

Figure 4.15: Solution of Linear Equations by the Gauss-Seidel Method (cont.)

```
          ERROR = .FALSE.
5         WRITE(OUT, 101)
          READ(IN,102) LAMBDA
          IF ((LAMBDA .LT. 0.0).OR.(LAMBDA .GT. 2.0)) GOTO 5
          N = NCOL
          DO 20 I = 1, N
            DO 10 J = 1, N
              B(I,J) = A(I,J)
10          CONTINUE
            W(I) = Y(I)
20        CONTINUE
          N1 = N - 1
          DO 70 I = 1, N1
            BIG = ABS(B(I,I))
            L = I
            I1 = I + 1
            DO 30 J = I1, N
              AB = ABS(B(J,I))
              IF (AB .LE. BIG) GOTO 30
                BIG = AB
                L = J
30          CONTINUE
            IF (BIG .EQ. 0.0) GOTO 99
            IF (L .EQ. I) GOTO 70
C -- Interchange rows to put largest element on diagonal.
            DO 40 J = 1, N
              CALL SWAP(B(L,J), B(I,J))
40          CONTINUE
            CALL SWAP(W(I), W(L))
70        CONTINUE
          IF (B(N,N) .EQ. 0.0) GOTO 99
C -- Initial values.
          DO 75 I = 1, N
          COEF(I) = 0.0
75        CONTINUE
          I = 0
80        I = I + 1
          DONE = .TRUE.
          DO 90 J = 1, N
            SUM = Y(J)
            DO 85 K = 1, N
              IF (J .NE. K) SUM = SUM - B(J,K) * COEF(K)
85          CONTINUE
            NEXTC = SUM / B(J,J)
            IF (ABS(NEXTC - COEF(J)) .LT. TOL) GOTO 95
            DONE = .FALSE.
            IF (NEXTC * COEF(J) .LT. 0.0)
     *        NEXTC = (COEF(J) + NEXTC) / 2
            COEF(J) = LAMBDA * NEXTC + (1 - LAMBDA)*COEF(J)
            WRITE(OUT, 103) I, J, COEF(J)
90        CONTINUE
```

Figure 4.15: Solution of Linear Equations by the Gauss-Seidel Method (cont.)

```
              IF ((I .LE. 100).AND.(.NOT. DONE)) GOTO 80
              WRITE(OUT, 998)
              ERROR = .TRUE.
    95        RETURN
    99        WRITE(OUT, 999)
              ERROR = .TRUE.
              RETURN
   101        FORMAT(' Relaxation factor? ')
   102        FORMAT(E10.0)
   103        FORMAT(I4, '; COEF(', I3, ') =', 1PE12.4)
   998        FORMAT(' No solution')
   999        FORMAT(' ERROR--Matrix singular')
              END
              SUBROUTINE SWAP(A,B)
C --  Interchange two values.
              REAL A, B, HOLD
C
              HOLD = A
              A = B
              B = HOLD
              RETURN
              END
              SUBROUTINE OUTPUT(A, Y, COEF, N)
C
C --  Print out the answers.
C
              INTEGER N, IN, OUT, I, J
              REAL A(8,8), Y(8), COEF(8)
              COMMON /INOUT/ IN, OUT
C
              DO 10 I = 1, N
                 WRITE(OUT, 101) (A(I,J), J = 1, N), Y(I)
    10        CONTINUE
              WRITE(OUT, 102)
              WRITE(OUT, 101) (COEF(I), I = 1, N)
              RETURN
   101        FORMAT(1P6E12.4)
   102        FORMAT(/ '    Solution')
              END
```

Figure 4.15: Solution of Linear Equations by the Gauss-Seidel Method (cont.)

Running the Gauss-Seidel Program

Generate a copy of Figure 4.15 and execute it. You will be asked for the number of equations. Enter the value of 3. Then enter the three equations from the electric circuit shown in Figure 4.1. The order of the equations is now immaterial since we have incorporated a row-interchange routine. You will next be asked for the relaxation factor. Give a value of 1.0. Each successive iteration will be printed out after the iteration number.

Convergence will occur after about 20 iterations. The program will again ask for the number of equations. Give a value of −3 this time. The minus sign indicates that the equations from the previous step are to be reused.

You can repeatedly rerun the program, trying different values for the relaxation factor. The following table shows the dependence of number of iterations on the choice of the relaxation factor.

Lambda	Iterations
0.8	31
1.0	20
1.2	11
1.3	9
1.4	12
1.5	15
1.8	44

Next, enter the 2-by-2 Hilbert matrix:

$$\begin{bmatrix} 1.0 & 0.5 \\ 0.5 & 0.3333 \end{bmatrix} \quad \begin{bmatrix} 1.5 \\ 0.83333 \end{bmatrix}$$

You will find that the optimum relaxation factor occurs at a value of 1.4, about the same as for the previous set of equations. Yet, for other sets of equations, the optimum value of lambda might be 1.0 or 0.8.

Clearly, the Gauss-Seidel is not as automatic a technique as the others we have considered. But it should be considered for solving large numbers of linear equations, or for solving sets of nonlinear equations.

SUMMARY

We have studied several methods for solving simultaneous equations, each method suited to a different situation. We have presented FORTRAN programs to carry out the algorithms of each of these methods. We have also investigated several special cases: multiple constant vectors; ill-conditioned equations; best-fit solutions for an "overdetermined" equation system; and equations with complex coefficients. In the programs for these special cases we have seen an abundance of new and powerful features of FORTRAN programming.

EXERCISES

4-1: *Show that the solution to the following set of linear equations:*

$$
\begin{aligned}
4x_1 + 3x_2 + 2x_3 + x_4 &= 10 \\
-1x_1 + 4x_2 - 2x_3 - x_4 &= 0 \\
- 2x_2 + 4x_3 + 3x_4 &= 5 \\
-2x_1 - x_2 + x_3 + 4x_4 &= 2
\end{aligned}
$$

is [1 1 1 1].

4-2: *Show that the solution to the equations:*

$$
\begin{aligned}
5x_1 + 2x_2 + x_3 + 3x_4 &= 6 \\
4x_1 + 3x_2 + x_3 - 2x_4 &= -4 \\
3x_1 + x_2 + 2x_3 + x_4 &= 2 \\
-x_1 + x_2 + 5x_3 + 3x_4 &= 7
\end{aligned}
$$

is [2 −3 −1 1].

4-3: *Interchange the inductor and the capacitor in Figure 4.13 and find the new loop currents.*

Answer:

$$
\begin{aligned}
I_1 &= 2.39 + j1.71 = 2.9\underline{/35.6^\circ} \\
I_2 &= 1.87 + j0.117 = 1.87\underline{/3.58^\circ}
\end{aligned}
$$

5

Development of a Curve-Fitting Program

IN THIS CHAPTER we will develop a least-squares curve-fitting program. In particular, we will develop a computer program for finding the straight line that can best represent a set of x-y data. This program will generate the data, calculate the desired equation, print out the results, produce a plot of the data, and supply a measure of the correlation between x and y. A sorting routine will be added in Chapter 6 to allow handling of real experimental data.

Although the resulting program will be large and complex, we will not program all the parts at one time. Rather, we will use a modular, top-down approach. Only a small portion of the program will be written at first. This part will be checked by actually running it. Another portion of the code will then be added, and it, too, will be checked by running the new program. In this way, a relatively large program can be developed in a logical fashion. Each step will be tested along the way. If an error appears, it will most likely be found in the most recently added portion.

As we develop the different parts of this program, we will be discussing several algorithms and their implementations. These include:

- the use of a random function and a "fudge factor" to simulate scattered-line experimental data

- a subroutine for plotting graphs on a regular character-oriented terminal or on a line printer

- a least-squares curve fitting routine, using differential calculus to arrive at the slope and y-intercept of the actual fitted data

- a simple but elegant method for integrating the correlation coefficient into our program.

THE MAIN PROGRAM

The first thing we will do is write the main program with the input and output routines. The main program will always contain as little as possible: the dimension statement and the calls to the various subroutines.

First Version: Input and Output Subroutines

Create the FORTRAN source program shown in Figure 5.1. Use a descriptive file name such as:

CFIT1.FOR

for the first version. The program begins by declaring the necessary identifiers and the dimensions of the vectors X and Y. Two subroutines are called. One provides the input and the other performs the output. A random number generator is called by the input subroutine. You may have to change the argument to, or the spelling of, function RAN if your FORTRAN contains a built-in random number generator. On the other hand, if your FORTRAN does not include such a function, you can use the one given in Figure 2.5.

```
C       PROGRAM CFIT1
C
C -- Linear least-squares fit
C -- May  1, 81
C
```

Figure 5.1: The Beginning of a Curve-Fitting Program

```
        INTEGER LENGTH, IN, OUT, MAX
        REAL X(20), Y(20), A, B
        COMMON /INOUT/ IN, OUT, MAX
C
        IN = 1
        OUT = 1
        MAX = 20
10      CALL INPUT(X, Y, LENGTH)
        CALL OUTPUT(X, Y, LENGTH)
        GOTO 10
        END
        SUBROUTINE INPUT(X, Y, N)
C -- Get values for N and arrays X and Y.
C
        INTEGER N, IN, OUT, I, J, MAX
        REAL X(1), Y(1), FUDGE
        COMMON /INOUT/ IN, OUT, MAX
        DATA A/2.0/, B/5.0/
C
        WRITE(OUT, 101)
        READ(IN,103) FUDGE
        IF (FUDGE .LT. 0.0) GOTO 99
5       WRITE(OUT, 102)
        READ(IN,104) N
        IF (N .GT. MAX) GOTO 5
        IF (N .LT. 0) GOTO 99
        DO 10 I = 1, N
          J = N + 1 - I
          X(I) = J
          Y(I) = (A + B*J)*(1.0+(2.0*RAN(0) - 1.0)*FUDGE)
10      CONTINUE
        RETURN
99      STOP
101     FORMAT(/' Fudge? ')
102     FORMAT(' How many points? ')
103     FORMAT(F10.0)
104     FORMAT(I2)
        END
        SUBROUTINE OUTPUT(X, Y, N)
C
C -- Print out the answers.
C
        INTEGER N, IN, OUT, I
        REAL X(1), Y(1)
        COMMON /INOUT/ IN, OUT
C
        WRITE(OUT, 101)
        WRITE(OUT, 102) (I, X(I), Y(I), I = 1, N)
        RETURN
101     FORMAT('    I       X         Y')
102     FORMAT(I4, F8.1, F9.2)
        END
```

Figure 5.1: The Beginning of a Curve-Fitting Program (cont.)

The input subroutine will initially generate a set of x-y points using a random number generator. In a later chapter we will alter this routine to read data embedded directly in the program. Another possibility is to read the data from a disk file.

Let us now look more closely at the input subroutine and the algorithm for simulating experimental data.

The Scattering Algorithm

The input subroutine generates a straight line with an intercept (A) of 2 and a slope (B) of 5. That is, it generates a set of data in the arrays X and Y, corresponding to the line:

$$y = 2 + 5x$$

A random number generator is then used to move the points off the line according to the variable FUDGE. If the value of FUDGE is zero, a perfectly straight line is generated. On the other hand, if FUDGE has a value of 0.2, the points will be displaced to a maximum of 20% from the line.

It would be more realistic to generate Gaussian random numbers than uniformly distributed random numbers in this application. However, our chosen method is much faster. Furthermore, we will be removing this part of the program when real data are incorporated into the curve-fitting routine.

The scattering algorithm works in the following way: function RAN returns a real number between 0 and 1. This value is doubled to give a range of 0 to 2. The subtraction of 1 sets the range from -1 to $+1$. Finally, this result is multiplied by the fudge factor and added to unity to give the desired range.

Running the Main Program

Compile the program and try it out. You will be asked to input a value for the fudge factor. Give a value of zero the first time. Then you will be asked for the number of points. Give the value of 9. The result will be three columns of numbers, as shown in Figure 5.2. Notice that the elements of the array X are generated in descending order. We will be able to reorder the arrays X and Y when we add a sorting routine in the next chapter.

At the completion of this task, the program asks for another value for the fudge factor. This time, respond with 0.2. The resulting x values should be the same as the previous values. The y values, however, will be somewhat larger or smaller than their previous values. When this step is finished, you will be asked for another value for the fudge factor. Enter a negative number this time to terminate the program.

```
Fudge? 0
How many points? 9

   I        X          Y
   1       9.0       47.00
   2       8.0       42.00
   3       7.0       37.00
   4       6.0       32.00
   5       5.0       27.00
   6       4.0       22.00
   7       3.0       17.00
   8       2.0       12.00
   9       1.0        7.00
```

Figure 5.2: Output:
First Run of the Curve-Fitting Program with the Fudge Factor Zero

Now that you have a working program, you can begin to add new features. But always keep copies of previous versions of a program. Then, if you have trouble with a new version, you can return to the previous version and start again.

A PRINTER PLOTTER ROUTINE

Next we will add a routine for plotting the results on an ordinary computer terminal. Large computers are typically provided with a digital plotter and perhaps a graphic video terminal. With these devices, it is possible to display data to a high degree of precision. Unfortunately, this approach is usually time-consuming. Furthermore, small computers may not have such devices.

As an alternative, experimental data can be displayed on a regular character-oriented terminal or on a line printer by using plus (+) and asterisk (*) symbols. The resulting plot will be a crude representation of the actual data. However, this plot may be useful for finding gross errors in programming as well as incorrectly entered data. Furthermore, the resulting plot can be viewed immediately along with the computational results, whereas there may be a delay in obtaining plotter output from a large computer.

The FORTRAN subroutine for plotting one or two dependent variables as a function of a third independent variable is shown in Figure 5.3. This subroutine plots the independent variable vertically rather than in the usual horizontal direction. The dependent variables are then displayed horizontally. That is, the graph is rotated clockwise a quarter turn from the usual axis orientation. We will use this plot to display uniformly spaced simple functions. However, multivalued functions can also be displayed. Furthermore, the values of the independent variable need not be uniformly spaced.

```
        SUBROUTINE PLOT(X,Y,YCALC,N,LUN,ILINES)
C
C -- May  1, 81
C------------------------------------------------------------------
C   Purpose--
C     Produce a plot on the printer of one or two vectors, e.g.,
C     Y and YCALC, as a function of a third vector, e.g., X.
C     Arrays must be sorted in ascending or descending order of X.
C
C   Description of parameters
C     X --      Vector of independent variables
C     Y --      Vector of dependent variables
C     YCALC--   Vector of second dependent variable
C     N --      Length of vectors. N is negative if only one
C               dependent variable.
C     LUN-      Logical unit number for output
C     ILINES -  Number of lines for plot
C------------------------------------------------------------------
C
        REAL X(1), Y(1), YCALC(1), YLABEL(6)
        INTEGER     OUT(51), PLUS, STAR, BLANK, MULT
        DATA IPLUS/1H+/, STAR/1H*/, BLANK/1H /, MULT/1HM/, LINEL/51/
        JF(P) = IFIX((P-YMIN)/YSCALE+1.0)
C
        LINES = MAXO(ILINES, MINO(N, 200))
        LONE = 0
        PLUS = IPLUS
        IF (N.GT.0) GOTO 2
C -- N is negative, plot only X and Y.
        N = -N
        LONE = 1
        PLUS = STAR
        WRITE(LUN,105) N, PLUS, MULT
2       XLOW = X(1)
        IF (LONE.EQ.0) WRITE(LUN,101) N, PLUS, STAR, MULT
        XHI = X(N)
        YMAX = Y(1)
        YMIN = YMAX
        XSCALE = (XHI-XLOW) / (LINES - 1)
        IF (LONE.EQ.1) GOTO 5
        DO 3 I = 1,N
           YMIN = AMIN1(YMIN, Y(I), YCALC(I))
           YMAX = AMAX1(YMAX, Y(I), YCALC(I))
```

Figure 5.3: A Plotting Subroutine

```
3         CONTINUE
          GOTO 7
5         DO 6 I = 1,N
            YMIN = AMIN1(YMIN, Y(I))
            YMAX = AMAX1(YMAX, Y(I))
6         CONTINUE
7         YSCALE = (YMAX - YMIN) / 50
          YS10 = YSCALE * 10.0
          YLABEL(1) = YMIN
          DO 9  KN = 1,4
            YLABEL(KN+1) = YLABEL(KN)+YS10
9         CONTINUE
          YLABEL(6) = YMAX
          IPRINT = 0
          IF (ABS(XHI).GE.1.E5.OR. ABS(XHI).LT.1.E-2)  IPRINT = 1
          DO 10 I = 1, LINEL
            OUT(I) = BLANK
10        CONTINUE
C -- First line.
          JP = JF(Y(1))
          OUT(JP) = PLUS
          IF (LONE.EQ.1) GOTO 12
          JP = JF(YCALC(1))
          OUT(JP) = STAR
12        L = 1
          XLABEL = XLOW
          ISKIP = 0
          DO 80 I = 2, LINES
          XNEXT = XLOW + XSCALE * (I-1)
          IF (ISKIP.EQ.1) GOTO 25
13        L = L+1
          CHANGE = XNEXT - 0.5 * XSCALE
          IF ((X(L) - CHANGE) * SIGN(1.0,XSCALE).GT.0.0) GOTO 30
C -- Multiple point.
          JP = JF(Y(L))
          IF ((OUT(JP).EQ.PLUS) .OR. (OUT(JP).EQ.MULT)) GOTO 15
          OUT(JP) = PLUS
          GOTO 20
15        OUT(JP) = MULT
20        IF (LONE .EQ. 1) GOTO 13
          JP = JF(YCALC(L))
          OUT(JP) = STAR
          GOTO 13
C -- Skip line.
25        WRITE(LUN,103)
          GOTO 40
C -- Print line.
30        DO 31 IX = 1, LINEL
            JX = LINEL - IX + 1
            IF (OUT(JX) .NE. BLANK) GOTO 32
31        CONTINUE
          JX = 1
```

Figure 5.3: A Plotting Subroutine (cont.)

```
32      IF (IPRINT.EQ.0) WRITE (LUN,104) XLABEL,(OUT(IX),IX = 1,JX)
        IF (IPRINT.EQ.1) WRITE (LUN,102) XLABEL,(OUT(IX),IX = 1,JX)
        DO 35 IX = 1, LINEL
          OUT(IX) = BLANK
35      CONTINUE
40      CHANGE = XNEXT + 0.5 * XSCALE
        IF ((X(L) - CHANGE) * SIGN(1.0,XSCALE) .LT. 0.0) GOTO 50
        ISKIP = 1
        GOTO 80
50      ISKIP = 0
        XLABEL = XNEXT
        JP = JF(Y(L))
        OUT(JP) = PLUS
        IF (LONE .EQ. 1) GOTO 80
        JP = JF(YCALC(L))
        OUT(JP) = STAR
80      CONTINUE
81      IF (L .GE. N) GOTO 90
C -- Multiple point for last line.
        L  = L+1
        JP = JF(Y(L))
        IF ((OUT(JP) .EQ. PLUS) .OR. (OUT(JP) .EQ. MULT)) GOTO 83
        OUT(JP) = PLUS
        GOTO 81
83      OUT(JP) = MULT
        GOTO 81
90      DO 91 IX = 1, LINEL
        JX = LINEL - IX + 1
        IF (OUT(JX) .NE. BLANK) GOTO 92
91      CONTINUE
        JX = 1
92      IF (IPRINT .EQ. 0) WRITE(LUN,104) XHI,(OUT(IX),IX = 1,JX)
        IF (IPRINT .EQ. 1) WRITE(LUN,102) XHI,(OUT(IX),IX = 1,JX)
        WRITE (LUN,107)
        IF ((ABS(YMAX).LT.1.E4).AND.(ABS(YMAX).GE.1.E-2)) GOTO 96
        WRITE (LUN,106) YLABEL
        GOTO 97
96      WRITE (LUN,108) YLABEL
97      WRITE (LUN,109)
        RETURN
101     FORMAT('0', I3, ' DATA SETS', 6X, A1,
      * '-DATA, ', A1, '-FITTED CURVE, ', A1, '-MULTIPLE POINT'/)
102     FORMAT(1X ,1PE11.4, 5X, 101A1)
103     FORMAT('      -')
104     FORMAT(1X ,F11.4, 5X, 101A1)
105     FORMAT('0', I3, ' DATA SETS', 6X,
      * A1, '-DATA, ', A1, '-MULTIPLE POINT'/)
106     FORMAT('0', 8X, 1P11E10.2)
107     FORMAT(17X, 5('^', 9X), '^')
108     FORMAT('0', 8X, 11F10.3)
109     FORMAT(/ 30X, 'DEPENDENT VARIABLE')
        END
```

Figure 5.3: A Plotting Subroutine (cont.)

Running the Plotter Routine

Make a copy of the previous version of your source program and give it a distinctive name such as CFIT2.FOR. Add the plotting subroutine to the end of the program. Alternatively, the plotting program can be placed into a separate file. It can be independently compiled, then linked to the main program. You must also, in either case, insert into the main program a call to the plotting subroutine. The new main program is given in Figure 5.4.

```
C        PROGRAM CFIT2
C
C -- Linear least-squares fit
C -- May  1, 81
C
         INTEGER LENGTH, IN, OUT, MAX, LINES
         REAL X(20), Y(20), A, B
         COMMON /INOUT/ IN, OUT, MAX
C
         IN = 1
         OUT = 1
         MAX = 20
10       CALL INPUT(X, Y, LENGTH)
         CALL OUTPUT(X, Y, LENGTH)
         LINES = 2*(LENGTH - 1) + 1
         CALL PLOT(X, Y, Y, -LENGTH, OUT, LINES)
         GOTO 10
         END
```

Figure 5.4: A call to the plotting routine is added.

The third parameter in the call to the plotting subroutine is the length of the vectors. Its value is negative, indicating that only a single function is to be plotted. The next parameter is the logical unit number for the plotter output. The final parameter designates the number of lines for the plot.

Compile this new version and try it out. Give a value of 0.2 for the fudge factor. The plotter output will look like Figure 5.5.

We are going to use the plotting routine with all of the remaining versions of the curve-fitting program in this chapter. Since this is such a large subroutine, you might want to place it into a separate disk file.

To test all aspects of our plotter routine, we must try plotting two curves at once. To do this, we will incorporate a small subroutine called LINFIT into our program in the next section. This subroutine will be replaced by a much more significant subroutine later in this chapter, after we have studied the least-squares curve-fitting algorithm.

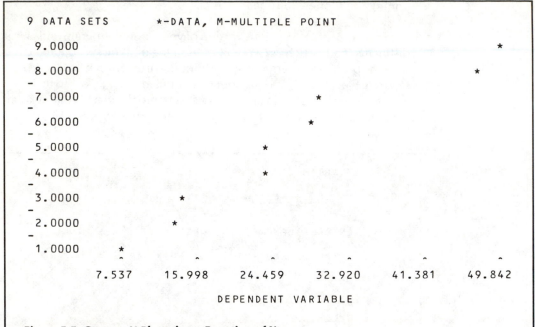

```
9 DATA SETS        *-DATA, M-MULTIPLE POINT
  9.0000                                                            *
  -
  8.0000                                                      *
  -
  7.0000                              *
  -
  6.0000                          *
  -
  5.0000                  *
  -
  4.0000                  *
  -
  3.0000          *
  -
  2.0000          *
  -
  1.0000      *
          ^       ^       ^       ^       ^       ^
        7.537   15.998  24.459  32.920  41.381  49.842
                   DEPENDENT VARIABLE
```

Figure 5.5: Output: Y Plotted as a Function of X

A SIMULATED CURVE FIT

For our next step, then, we will add a simulated curve-fitting sub-routine. This routine will simply generate a vector YCALC that lies on our straight line with a slope of 5 and an intercept of 2. Keep in mind that we will add a proper curve-fitting routine in a later version.

Alter the main program so that it looks like Figure 5.6, and change the output subroutine to look like Figure 5.7. Add subroutine LINFIT, shown in Figure 5.8, to your source program.

```
C        PROGRAM CFIT3
C
C -- Linear least-squares fit
C -- May  1, 81
C
         INTEGER LENGTH, IN, OUT, MAX, LINES
         REAL X(20), Y(20), A, B, YCALC(20)
         COMMON /INOUT/ IN, OUT, MAX
C
         IN = 1
         OUT = 1
         MAX = 20
```

Figure 5.6: The Current Main Program

```
10      CALL INPUT(X, Y, LENGTH)
        CALL LINFIT(X, Y, YCALC, A, B, LENGTH)
        CALL OUTPUT(X, Y, YCALC, LENGTH)
        LINES = 2 * (LENGTH - 1) + 1
        CALL PLOT(X, Y, YCALC, LENGTH, OUT, LINES)
        GOTO 10
        END
```

Figure 5.6: The Current Main Program (cont.)

```
        SUBROUTINE OUTPUT(X, Y, YCALC, N)
C
C -- Print out the answers.
C
        INTEGER N, IN, OUT, I
        REAL X(1), Y(1), YCALC(1)
        COMMON /INOUT/ IN, OUT
C
        WRITE(OUT, 101)
        WRITE(OUT, 102) (I, X(I), Y(I), YCALC(I), I = 1, N)
        RETURN
101     FORMAT('   I       X        Y      Y CALC')
102     FORMAT(I4, F8.1, 2F9.2)
        END
```

Figure 5.7: The Revised Output Routine

```
        SUBROUTINE LINFIT(X, Y, YCALC, A, B, N)
C
C -- Generate a straight line for X-Y.
C
        INTEGER N, I
        REAL X(1), Y(1), YCALC(1), A, B
C
        A = 2
        B = 5
        DO 10 I = 1, N
          YCALC(I) = A + B * X(I)
10      CONTINUE
        RETURN
        END
```

Figure 5.8: Subroutine LINFIT to Simulate a Linear Fit

Running the Plotter Routine with Two Curves

Compile this new version and try it out. Again, you will be asked to give a value for the fudge factor. Respond with an answer of 0.2. This will produce four columns of data, including the values for YCALC. The printer plot that follows the tabular data will show two different sets of data. Asterisks (*) are used to represent the values of YCALC. They will form a nearly perfect straight line. In addition, plus symbols (+) represent the values of Y. They will be scattered on both sides of the calculated values (YCALC) as shown in Figure 5.9. If two values of Y are the same, an M symbol is printed. However, if the values of Y and YCALC are coincident for a given value of X, only the asterisk is given.

When the program asks for another value for the fudge factor, respond with a zero. For this example, Y and YCALC will both have the same values. Since the two lines lie one on top of the other, only one line of asterisks will be shown. Finally, give a negative value for the fudge factor to terminate the program.

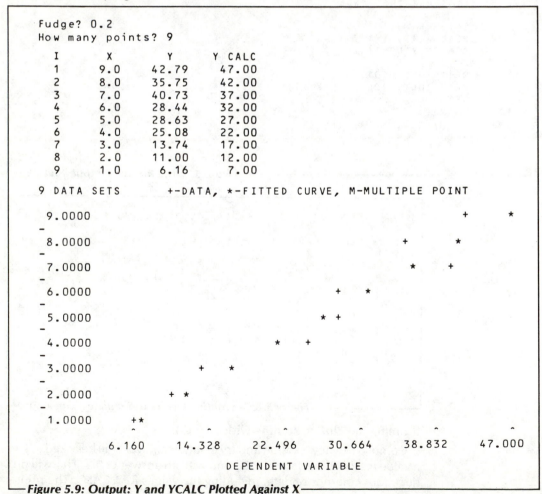

```
 Fudge? 0.2
 How many points? 9

      I        X        Y      Y CALC
      1      9.0     42.79     47.00
      2      8.0     35.75     42.00
      3      7.0     40.73     37.00
      4      6.0     28.44     32.00
      5      5.0     28.63     27.00
      6      4.0     25.08     22.00
      7      3.0     13.74     17.00
      8      2.0     11.00     12.00
      9      1.0      6.16      7.00

 9 DATA SETS      +-DATA, *-FITTED CURVE, M-MULTIPLE POINT

  9.0000                                              +        *
  -
  8.0000                                         +        *
  -
  7.0000                                       *        +
  -
  6.0000                                  +        *
  -
  5.0000                                *  +
  -
  4.0000                          *        +
  -
  3.0000                  +        *
  -
  2.0000              +  *
  -
  1.0000        +*
              ^         ^         ^         ^         ^         ^
           6.160    14.328    22.496    30.664    38.832    47.000
                      DEPENDENT VARIABLE
```

Figure 5.9: Output: Y and YCALC Plotted Against X

Now that we have written and tested both the main program and a procedure for plotting curves, we are ready to move on to the real topic of this chapter.

THE CURVE-FITTING ALGORITHM

Though it may appear from Figure 5.9 that we have completed our curve-fitting program, we have actually not done so. We have simply generated a set of data (YCALC) that corresponds to our original line. The time has come to derive the algorithm for a linear, least-squares procedure. We will first introduce a new vector, **r**, which contains the *residuals*.

For each experimental point, corresponding to an x-y pair, there will be an element of **r** that represents the difference between the corresponding calculated value $\hat{y}$ (pronounced y-hat) and the original value of y. This can be expressed mathematically as:

$$r_i = \hat{y}_i - y_i \tag{1}$$

Occasionally, a point will coincide with the calculated curve. But in general, about half of the x-y points will lie on one side of the fitted curve, resulting in a positive value for r. The remaining points will lie on the other side of the curve and give negative values for r. The sum of these residuals should be close to zero.

The least-squares, curve-fitting criterion is that the sum of the squares of the residuals be minimized. The square of each residual will be positive; therefore, the sum of the residuals squared (*SRS*) will be a positive number. This criterion can be expressed as:

$$SRS = \sum_{i=1}^{n} r_i^2 = \text{minimum} \tag{2}$$

Where n is the number of x-y points (and the length of the vectors **x**, **y**, and **ŷ**).

By combining Equation 1 with the curve-fitting equation:

$$\hat{y}_i = A + Bx_i \tag{3}$$

we get

$$r_i = A + Bx_i - y_i \tag{4}$$

and

$$SRS = \sum_{i=1}^{n} r_i^2 = \sum_{i=1}^{n} (A + Bx_i - y_i)^2 \tag{5}$$

The problem is reduced to finding the values of A and B so that the summation of Equation 5 is minimized. This is accomplished with differential calculus. We take the derivative of Equation 5 with respect to each variable (A and B in this case) and set the result to zero.

$$\frac{\delta \Sigma r_i^2}{\delta A} = 0 \quad \text{and} \quad \frac{\delta \Sigma r_i^2}{\delta B} = 0 \tag{6}$$

Substitution of Equation 5 into Equations 6 gives:

$$\frac{\delta \Sigma (A + Bx_i - y_i)^2}{\delta A} = 0 \tag{7}$$

and

$$\frac{\delta \Sigma (A + Bx_i - y_i)^2}{\delta B} = 0 \tag{8}$$

which is equivalent to:

$$\frac{2 \Sigma (A + Bx_i - y_i) \, \delta \Sigma (A + Bx_i - y_i)}{\delta A} = 0 \tag{9}$$

and

$$\frac{2 \Sigma (A + Bx_i - y_i) \, \delta \Sigma (A + Bx_i - y_i)}{\delta B} = 0 \tag{10}$$

Since B, x, and y are not functions of A, and the derivative of A with respect to itself is unity, Equation 9 reduces to:

$$\Sigma A + \Sigma Bx_i = \Sigma y_i \tag{11}$$

Similarly, A, x and y are not functions of B. Therefore, Equation 10 becomes:

$$\Sigma Ax_i + \Sigma Bx_i^2 = \Sigma x_i y_i \tag{12}$$

A and B are constants. Therefore they can be factored from the summation step. Equations 7 and 8 can then be expressed as:

$$An + B\Sigma x_i = \Sigma y_i \tag{13}$$

and

$$A\Sigma x_i + B\Sigma x_i^2 = \Sigma x_i y_i \tag{14}$$

We have thus reduced the problem of finding a straight line through a set of x-y data points to one of solving two simultaneous equations (13 and 14). Both of these equations are linear in the unknowns A and B. (x, y, and n are, of course, the original data.) The simultaneous solution can be obtained by using Cramer's rule:

$$A = \frac{\begin{vmatrix} \Sigma y_i & \Sigma x_i \\ \Sigma x_i y_i & \Sigma x_i^2 \end{vmatrix}}{\begin{vmatrix} n & \Sigma x_i \\ \Sigma x_i & \Sigma x_i^2 \end{vmatrix}} \tag{15}$$

and

$$B = \frac{\begin{vmatrix} n & \Sigma y_i \\ \Sigma x_i & \Sigma x_i y_i \end{vmatrix}}{\begin{vmatrix} n & \Sigma x_i \\ \Sigma x_i & \Sigma x_i^2 \end{vmatrix}} \tag{16}$$

The corresponding equations we have to solve are:

$$A = \frac{\Sigma x_i^2 \Sigma y_i - \Sigma x_i \Sigma x_i y_i}{n \Sigma x_i^2 - \Sigma x_i \Sigma x_i} \tag{17}$$

and

$$B = \frac{n \Sigma x_i y_i - \Sigma x_i \Sigma y_i}{n \Sigma x_i^2 - \Sigma x_i \Sigma x_i} \tag{18}$$

The computer calculation of A and B is straightforward. The summation of x is obtained by summing the values of the array X. The summation of x^2 is obtained by squaring each value of X and then adding up the squares.

Equations 17 and 18 are commonly converted into an equivalent form by dividing the numerator and denominator by n:

$$A = \frac{(\Sigma x_i^2 \Sigma y_i - \Sigma x_i \Sigma x_i y_i)/n}{\Sigma x_i^2 - \Sigma x_i \Sigma x_i/n} \tag{19}$$

and

$$B = \frac{\Sigma x_i y_i - \Sigma x_i \Sigma y_i/n}{\Sigma x_i^2 - \Sigma x_i \Sigma x_i/n} \tag{20}$$

The denominators of Equations 19 and 20 appear in the formula for the standard deviation discussed in Chapter 2.

We have outlined the mathematics of least-squares curve-fitting and have derived formulas for finding the slope (B) and y-intercept (A) of a linear fitted curve. Now we are ready to add this feature to our program.

The Curve-Fitting Procedure

A program for fitting a straight line is shown in Figure 5.10. Make a copy of the previous program (Figure 5.6). Then alter the main program and subroutines OUTPUT and LINFIT (Figures 5.7 and 5.8) so they look like the version shown in Figure 5.10.

At the beginning of the new version of subroutine LINFIT the variables SUMX, SUMY, SUMXY, SUMX2, and SUMY2 are set to zero. These

```
C         PROGRAM CFIT4
C
C -- Linear least-squares fit
C -- May  1, 81
C
          INTEGER LENGTH, IN, OUT, MAX, LINES
          REAL X(20), Y(20), A, B, YCALC(20)
          COMMON /INOUT/ IN, OUT, MAX
C
          IN = 1
          OUT = 1
          MAX = 20
10        CALL INPUT(X, Y, LENGTH)
          CALL LINFIT(X, Y, YCALC, A, B, LENGTH)
          CALL OUTPUT(X, Y, YCALC, LENGTH, A, B)
          LINES = 2*(LENGTH - 1) + 1
          CALL PLOT(X, Y, YCALC, LENGTH, OUT, LINES)
          GOTO 10
          END
          SUBROUTINE INPUT(X, Y, N)
C -- Get values for N and arrays X and Y.
C
          INTEGER N, IN, OUT, I, J, MAX
          REAL X(1), Y(1), FUDGE
          COMMON /INOUT/ IN, OUT, MAX
          DATA A/2.0/, B/5.0/
C
          WRITE(OUT, 101)
          READ(IN,103) FUDGE
          IF (FUDGE .LT. 0.0) GOTO 99
5         WRITE(OUT, 102)
          READ(IN,104) N
          IF (N .GT. MAX) GOTO 5
          IF (N .LT. 0) GOTO 99
          DO 10 I = 1, N
            J = N + 1 - I
            X(I) = J
            Y(I) = (A + B*J)*(1.0+(2.0*RAN(0) - 1.0)*FUDGE)
10        CONTINUE
          RETURN
99        STOP
101       FORMAT(/' Fudge? ')
102       FORMAT(' How many points? ')
103       FORMAT(F10.0)
104       FORMAT(I2)
          END
```

Figure 5.10: Program to Generate a Least-Squares Fit

```
        SUBROUTINE OUTPUT(X, Y, YCALC, N, A, B)
C
C -- Print out the answers.
C
        INTEGER N, IN, OUT, I
        REAL X(1), Y(1), YCALC(1)
        COMMON /INOUT/ IN, OUT

C
        WRITE(OUT, 101)
        WRITE(OUT, 102) (I, X(I), Y(I), YCALC(I), I = 1, N)
        WRITE(OUT, 103) A
        WRITE(OUT, 104) B
        RETURN
101     FORMAT('  I         X          Y        Y CALC')
102     FORMAT(I4, F8.1, 2F9.2)
103     FORMAT(/' Intercept is', F7.2)
104     FORMAT('      Slope is', F7.2)
        END
        SUBROUTINE LINFIT(X, Y, YCALC, A, B, N)
C
C -- Fit a straight line for (YCALC) through
C -- N pairs of X-Y points.
C
        INTEGER N, I
        REAL X(1), Y(1), YCALC(1), A, B, SXY, XYY, SXX
        REAL SUMX, SUMY, SUMXY, SUMX2, SUMY2, XI, YI
C
        SUMX = 0.0
        SUMY = 0.0
        SUMXY = 0.0
        SUMX2 = 0.0
        SUMY2 = 0.0
        DO 10 I = 1, N
          XI = X(I)
          YI = Y(I)
          SUMX = SUMX + XI
          SUMY = SUMY + YI
          SUMXY = SUMXY + XI * YI
          SUMX2 = SUMX2 + XI * XI
          SUMY2 = SUMY2 + YI * YI
10      CONTINUE
        SXX = SUMX2 - SUMX * SUMX / N
        SXY = SUMXY - SUMX * SUMY / N
        SYY = SUMY2 - SUMY * SUMY / N
        B = SXY / SXX
        A = (SUMY - B * SUMX) / N
        DO 20 I = 1, N
          YCALC(I) = A + B * X(I)
20      CONTINUE
        RETURN
        END
```

Figure 5.10: Program to Generate a Least-Squares Fit (cont.)

variables accumulate the needed sums in the first DO loop. Notice that a change of variable is made at the beginning of the loop:

    XI = X(I)
    YI = Y(I)

It generally takes longer to access an element of an array than to access a scalar value. Furthermore, the vector X is a dummy variable. Therefore, the elements must be obtained from the calling program. By contrast, the scalar XI is a local variable. Consequently, when the same value of an array is needed many times in a loop, it will generally be faster to define a scalar value and use that instead. On the other hand, some modern compilers incorporate an *optimizer* that automatically performs this task. If this is so, the transformation is unnecessary.

Running the Curve-Fitting Program

Run the new version to try it out. Give a fudge factor of 0.2 at first. The fitted line of asterisks should go neatly through the scattered plus symbols. Compare Figure 5.11, which shows the actual fit, with Figure 5.9, which gives the simulated fit. Next, make the fudge factor equal to zero. Only a single line of asterisks should be apparent now. Furthermore, the intercept should be equal to 2 and the slope should be equal to 5, the initial values.

THE CORRELATION COEFFICIENT

Although we now have a procedure for calculating the desired straight line, we are not finished yet. We can obtain the equation of a line fitting our experimental data. Then we can use this equation to predict a value for y from a given value of x. Under certain circumstances, however, our mathematically correct solution is useless.

Consider, for example, the set of data shown in Figure 5.12. Our curve-fitting program can find the equation of a straight line through the data. But the resulting line does not give us any additional information. That is, a knowledge of the behavior of x does not tell us anything about the behavior of y. There is no correlation between x and y.

Figure 5.13 shows another case where a knowledge of x is no help in predicting the behavior of y. Again, there is no correlation between x and y.

We need to obtain a quantitative measure of the correlation between x and y. We want to know how well we can predict the behavior of y if we know the behavior of x. The measure we need is the *correlation coefficient*.

We saw in Chapter 2 that we could characterize a set of data by the

```
Fudge? 0.2
How many points? 9

    I         X          Y        Y CALC
    1        9.0        51.33      48.69
    2        8.0        35.46      43.60
    3        7.0        41.64      38.50
    4        6.0        34.46      33.40
    5        5.0        31.87      28.30
    6        4.0        23.16      23.21
    7        3.0        17.20      18.11
    8        2.0        12.42      13.01
    9        1.0         7.19       7.91

Intercept is    2.82
   Slope is     5.10

9 DATA SETS      +-DATA, *-FITTED CURVE, M-MULTIPLE POINT

   9.0000                                                  *   +
   -
   8.0000                                   +          *
   -
   7.0000                                        *    +
   -
   6.0000                                 *+
   -
   5.0000                          *    +
   -
   4.0000                     *
   -
   3.0000              +*
   -
   2.0000          +*
   -
   1.0000      *
            ^           ^           ^           ^           ^           ^
          7.190      16.018      24.845      33.673      42.501      51.328

                       DEPENDENT VARIABLE
```

Figure 5.11: Output: A Least-Squares Fit to y vs x

mean and the standard deviation of the values about their own mean. We can also calculate the standard deviation of y about the fitted curve. This measure is termed the *standard error of the estimate* (*SEE*). The correlation coefficient compares the standard deviation of y (about its own mean) to the standard deviation about the fitted curve (SEE).

The correlation coefficient is zero when there is no correlation. The data in Figures 5.12 and 5.13 are examples of this. On the other hand, the correlation coefficient approaches unity as the data approach a straight line. The correlation coefficient for the data given in Figure 5.11 is 0.99.

We will now incorporate calculations for the correlation coefficient into our program.

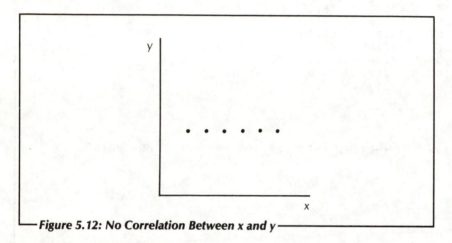

Figure 5.12: No Correlation Between x and y

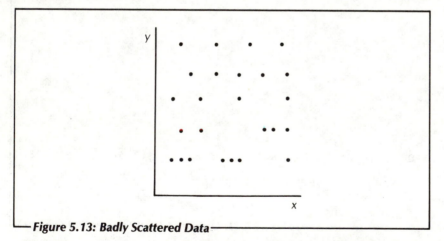

Figure 5.13: Badly Scattered Data

FORTRAN PROGRAM: LEAST-SQUARES CURVE FITTING FOR SIMULATED DATA

With a few additional statements, our program can calculate the correlation coefficient and the standard errors of the coefficients (the intercept and slope). The standard errors are standard deviations for the coefficients. They can be used to determine confidence intervals for each coefficient.

In Figure 5.14, subroutines LINFIT and OUTPUT and the main program have been altered so that the standard errors and the correlation coefficient are generated. With the correlation coefficient we have added a final test to our program—a way of quantifying the usefulness of the fitted curve. Let us look now at the finished program.

```
C       PROGRAM CFIT5
C
C -- Linear least-squares fit
C -- May  1, 81
C
        INTEGER LENGTH, IN, OUT, MAX, LINES
        REAL X(20), Y(20), A, B, YCALC(20), SIGMAA, SIGMAB, CORREL
        COMMON /INOUT/ IN, OUT, MAX
C
        IN = 1
        OUT = 1
        MAX = 20
10      CALL INPUT(X, Y, LENGTH)
        CALL LINFIT(X, Y, YCALC, A, B, LENGTH, SIGMAA, SIGMAB, CORREL)
        CALL OUTPUT(X, Y, YCALC, LENGTH, A, B, SIGMAA, SIGMAB, CORREL)
        LINES = 2*(LENGTH - 1) + 1
        CALL PLOT(X, Y, YCALC, LENGTH, OUT, LINES)
        GOTO 10
        END
        SUBROUTINE INPUT(X, Y, N)
C -- Get values for N and arrays X and Y.
C
        INTEGER N, IN, OUT, I, J, MAX
        REAL X(1), Y(1), FUDGE
        COMMON /INOUT/ IN, OUT, MAX
        DATA A/2.0/, B/5.0/
C
        WRITE(OUT, 101)
        READ(IN,103) FUDGE
        IF (FUDGE .LT. 0.0) GOTO 99
5       WRITE(OUT, 102)
        READ(IN,104) N
        IF (N .GT. MAX) GOTO 5
        IF (N .LT. 0) GOTO 99
        DO 10 I = 1, N
          J = N + 1 - I
          X(I) = J
          Y(I) = (A + B*J)*(1.0+(2.0*RAN(0) - 1.0)*FUDGE)
10      CONTINUE
        RETURN
99      STOP
101     FORMAT(/' Fudge? ')
102     FORMAT(' How many points? ')
```

Figure 5.14: The Complete Curve-Fitting Program

```
103     FORMAT(F10.0)
104     FORMAT(I2)
        END
        SUBROUTINE OUTPUT(X, Y, YCALC, N, A, B, SIGMAA, SIGMAB, CORREL)
C
C -- Print out the answers.
C
        INTEGER N, IN, OUT, I
        REAL X(1), Y(1), YCALC(1), A, B, SIGMAA, SIGMAB, CORREL
        COMMON /INOUT/ IN, OUT
C
        WRITE(OUT, 101)
        WRITE(OUT, 102) (I, X(I), Y(I), YCALC(I), I = 1, N)
        WRITE(OUT, 103) A, SIGMAA
        WRITE(OUT, 104) B, SIGMAB
        WRITE(OUT, 105) CORREL
        RETURN
101     FORMAT('   I        X            Y        Y CALC')
102     FORMAT(I4, F8.1, 2F9.2)
103     FORMAT(/' Intercept is', F7.2, ', sigma A is', F8.3)
104     FORMAT( '     Slope is', F7.2, ', sigma B is', F8.3)
105     FORMAT(/' Correlation coefficient is', F7.4)
        END
        SUBROUTINE LINFIT(X, Y, YCALC, A, B, N, SIGA, SIGB, COR)
C
C -- Fit a straight line for (YCALC) through
C -- N pairs of X-Y points.
C
        INTEGER N, I
        REAL X(1), Y(1), YCALC(1), A, B, SXY, XYY, SXX
        REAL SUMX, SUMY, SUMXY, SUMX2, SUMY2, XI, YI
        REAL SIGA, SIGB, COR, SEE
C
        SUMX = 0.0
        SUMY = 0.0
        SUMXY = 0.0
        SUMX2 = 0.0
        SUMY2 = 0.0
        DO 10 I = 1, N
           XI = X(I)
           YI = Y(I)
           SUMX = SUMX + XI
           SUMY = SUMY + YI
           SUMXY = SUMXY + XI * YI
           SUMX2 = SUMX2 + XI * XI
           SUMY2 = SUMY2 + YI * YI
10      CONTINUE
        SXX = SUMX2 - SUMX * SUMX / N
        SXY = SUMXY - SUMX * SUMY / N
        SYY = SUMY2 - SUMY * SUMY / N
        B = SXY / SXX
        A = (SUMY - B * SUMX) / N
```

Figure 5.14: The Complete Curve-Fitting Program (cont.)

```
        DO 20 I = 1, N
          YCALC(I) = A + B * X(I)
  20    CONTINUE
        COR = SXY / SQRT(SXX * SYY)
        SEE = SQRT((SUMY2 - A * SUMY - B * SUMXY) / (N - 2))
        SIGB = SEE / SQRT(SXX)
        SIGA = SIGB * SQRT(SUMX2 / N)
        RETURN
        END
```

Figure 5.14: The Complete Curve-Fitting Program (cont.)

Run the new version to try it out. First set the fudge factor to a value of zero. The intercept will be 2 and the slope will be 5, as before. In addition, the sigmas for A and B will be zero and the correlation coefficient will be equal to one. The plot will show a nearly straight line of asterisks. Now try a fudge factor of 0.2. This will give sigmas greater than zero for the intercept and the slope. The correlation coefficient will be somewhat less than unity.

SUMMARY

The modular development process that we have used for this program has allowed us to evaluate each subroutine as we wrote it. We began with the main program and added the subroutines step by step to plot the results, simulate data, compute the fitted curve, and supply the correlation co-efficient. Significantly enough, our final curve-fitting routine replaced an earlier version that we wrote solely for testing the plotting routine.

While we now have a linear curve-fitting program that works, it is not very useful. It can only fit data produced by a random number generator. Consequently, we will want to alter the input routine so that it can obtain actual data that is embedded in the program, or read from the keyboard or from a disk file. We will delay this step, however, until we add a sorting routine in the next chapter.

EXERCISES

5-1: *The vapor pressure of water is:*

temp, C	press, torr
20	17.535
30	31.824
40	55.324
50	92.51
60	149.38
70	233.7
80	355.1
90	525.76
100	760.00

where the temperature is given in degrees Celsius and the pressure is in torr (mm. of mercury). Find the coefficients A and B to the equation:

$$\ln P = A + B/T$$

by performing a least-squares fit on the data. Be sure to convert the temperature to Kelvin by adding 273.15 to the Celsius value. Change the input routine to generate the x values as the reciprocal of the temperature in Kelvin:

$$X(I) = 1.0 \,/\, ((I - 1)*10 + 20 + 273.15)$$

The y values can be read from a DATA statement or defined directly in the program, and then converted to the logarithm:

$$Y(I) = LOG(Y(I))$$

Answer: A = 20.46, B = −5153

6

Sorting

IN THIS CHAPTER we will develop several different sorting algorithms: two bubble sorts, a Shell sort, and a nonrecursive quick sort. Then we will incorporate one of these routines into the curve-fitting program developed in the previous chapter. First, let us discuss the rationale for including a sort routine in our program.

HANDLING EXPERIMENTAL DATA

The curve-fitting program we wrote in the previous chapter obtained its data from a random number generator. More realistic programs will obtain the data from the keyboard, or from a disk file. As an alternative, the data can be embedded in the program itself.

In particular, we might want to fit experimental data that have not been acquired in numeric order. As an example, suppose that the thermal expansion of a material is to be investigated. Paired measurements of temperature and the corresponding total length could be taken. The experimental apparatus is brought to a certain temperature and the length is measured. Then the temperature is increased and the sample temperature and length are measured again. However, it would not be wise to continue the experiment in this fashion. The problem is that a length might be measured before the sample temperature became uniform, or before the sample reached the value of the temperature sensor. The resulting measured pairs would all contain errors in the same direction.

A better experimental technique would be to approach the desired temperature sometimes from below and sometimes from above. For example, a collection of experimental data might look like the table below:

Temperature	Length
100	8.0
300	19.0
200	13.5
500	30.0
400	24.5

The curve-fitting program we developed in the last chapter could readily find a least-squares fit to this data. We could place the data directly in the input subroutine, as shown in Figure 6.1. Notice, however, that the scalar N and the vectors X and Y are dummy variables. Consequently, they cannot be initialized in a DATA statement.

```
SUBROUTINE INPUT(X, Y, N)
 . . .
N = 5
X(1) = 100
Y(1) =   8.0
X(2) = 300
Y(2) = 19.0
X(3) = 200
Y(3) = 13.5
X(4) = 500
Y(4) = 30.0
X(5) = 400
Y(5) = 24.5
 . . .
```

Figure 6.1: Experimental Data in an Input Subroutine

If we now run the program, however, we find that the plotting subroutine will give an incorrect rendering of the data. The problem is that the array of independent variables, X, must be arranged either in increasing or in decreasing order. That is, the data must be sorted. The plotting subroutine worked properly in the previous chapter because the X array was generated in decreasing order. Notice that we are not concerned about the ordering of the dependent variable Y.

We will begin our discussion of sorting routines with the easiest one to program—the bubble sort.

A BUBBLE SORT

To sort a collection of items is to arrange them in increasing or decreasing order. The items might be elements of an array of real numbers, or they might be a group of alphabetic and numeric characters called records. There are many different sorting algorithms. Some are very fast, others are very slow. Some are faster with nearly sorted data and some are slower under these conditions. Some require additional working space, others need only the space occupied by the original data.

The first sorting algorithm we will consider is known as the *bubble sort*. This routine is the easiest to understand and the easiest to program. Unfortunately, it is also the slowest. For sorting lists that contain fewer than a dozen or so items, however, the difference in speed is unimportant.

In the bubble sort, each element is compared to all the remaining, unsorted elements. If a particular pair is found to be out of order, the two elements are interchanged. There are two loops, one nested inside the other. The outer loop runs from 1 to one less than the length of the array. The inner loop runs from one larger than the outer loop up to the length of the array. With this algorithm, the smaller items "float" to the top of the array during the sorting process. This is of course the origin of the name "bubble sort".

We will write this first sorting routine as part of a complete program designed to test the relative efficiency of different sorting routines under different conditions. However, we will actually compile the sorting subroutine separately from the rest of the program. This step will make it relatively easy to substitute the rest of the sorting routines in this chapter.

FORTRAN PROGRAM: THE BUBBLE SORT

The program shown in Figure 6.2 contains a driver routine with its input subprogram. The bubble sort is given separately in Figure 6.3. The random number generator, RAN, from Figure 2.5 is also needed if your FORTRAN does not provide one. When the program is executed, it asks for the number of items to be sorted. The random number generator is then used to fill the vector X with the desired number of elements having values from zero to 100. The original set of numbers is printed on the console, ten per line. Statements such as:

```
WRITE(OUT, 101) BELL
```

sound the console bell at the beginning and end of the sorting process. This will allow a comparison to be made with the other sorting routines to be presented in this chapter. If you are not using an ASCII terminal, then these statements will have to be changed or removed.

At the end of the sorting step, the sorted array is printed along with the word RANDOM. The console bell sounds a third time, and the sorting subroutine is called again. This time, however, the sorting routine runs on an array that is already sorted. At the end of this second sorting step, the console bell sounds a fourth time and the sorted array is printed out again. The word SORTED is also printed. For the third phase of the test program, an array of numbers is generated in reverse order. The sorting routine is then called for the third time. At the end of the step, the sorted array is printed, along with the word REVERSED.

Type up the program given in Figure 6.2 and the sorting routine given in Figure 6.3. Separately compile each part and execute the combination. Be sure to include function RAN. Try to find a length that requires several minutes for sorting. Record the time needed for sorting each of the three arrangements of data. Then a comparison can be made with the other sorting routines given later in this chapter. If you are not using a video terminal, you will want to disable the printing of the arrays.

```
C       PROGRAM BSORT
C
C  -- Test sort routine.
C  -- Function RAN and subroutine SORT needed.
C  -- May  2, 81
C
        INTEGER LENGTH, IN, OUT, MAX, BELL, I
        REAL X(500)
        COMMON /INOUT/ IN, OUT, MAX
        DATA BELL/'G'/
C
        IN = 1
        OUT = 1
        MAX = 500
        BELL = BELL - 64
10      CALL INPUT(X, LENGTH)
        WRITE(OUT, 105) (X(I), I = 1, LENGTH)
        WRITE(OUT, 101) BELL
C  -- Random list
        CALL SORT(X, LENGTH)
        WRITE(OUT, 101) BELL
        WRITE(OUT, 105) (X(I), I = 1, LENGTH)
        WRITE(OUT, 102) BELL
C  -- Sorted list
        CALL SORT(X, LENGTH)
        WRITE(OUT, 101) BELL
        WRITE(OUT, 105) (X(I), I = 1, LENGTH)
        WRITE(OUT, 103) BELL
        DO 20 I = 1, LENGTH
          X(I) = LENGTH + 1 - I
20      CONTINUE
        WRITE(OUT, 101) BELL
C  -- Reversed list
        CALL SORT(X, LENGTH)
```

Figure 6.2: A Program to Test the Sorting Routine

```
          WRITE(OUT, 101) BELL
          WRITE(OUT, 105) (X(I), I = 1, LENGTH)
          WRITE(OUT, 104) BELL
          GOTO 10
101       FORMAT(1X, A4)
102       FORMAT(' Random', A4)
103       FORMAT(' Sorted', A4)
104       FORMAT(' Reversed', A4)
105       FORMAT(1X, 10F7.2)
          END
          SUBROUTINE INPUT(X, N)
C
C -- Get N and generate random vector X.
C
          INTEGER N, IN, OUT, I, J, MAX
          REAL X(1), Y(1)
          COMMON /INOUT/ IN, OUT, MAX
C
5         WRITE(OUT, 101)
          READ(IN,102) N
          IF (N .GT. MAX) GOTO 5
          IF (N .LT. 0) GOTO 99
          DO 10 I = 1, N
             X(I) = RAN(0) * 100.0
10        CONTINUE
          RETURN
99        STOP
101       FORMAT(' How many points? ')
102       FORMAT(I3)
          END
```

Figure 6.2: A Program to Test the Sorting Routine (cont.)

```
          SUBROUTINE SORT(A, N)
C
C -- Bubble sort routine for vector A.
C -- May  2, 81
C
          INTEGER N, I, J, N1, I1
          REAL A(1), HOLD
C
          N1 = N - 1
          DO 20 I = 1, N1
             I1 = I + 1
             DO 10 J = I1, N
                IF (A(I) .LE. A(J)) GOTO 10
                HOLD = A(I)
                A(I) = A(J)
                A(J) = HOLD
10           CONTINUE
20        CONTINUE
          RETURN
          END
```

Figure 6.3: Version One of the Bubble Sort Subroutine

We will now develop a slightly improved bubble sort routine, incorporating a swap function. The reason for this improvement will become clear as the chapter progresses.

Adding a Swap Function

Later in this chapter, we will incorporate a sorting procedure into the curve-fitting program from the previous chapter. At that time, we will want to interchange a value of a second array (Y) whenever we interchange the corresponding value of the first array (X). The middle part of the sort routine might then become:

```
HOLD = X(I)
X(I) = X(J)
X(J) = HOLD
HOLD = Y(I)
Y(I) = Y(J)
Y(J) = HOLD
```

In this example, first a pair of elements of one array is interchanged, and then the corresponding elements in the other array are swapped. As in Figure 6.2, the interchange operation requires a third variable, called HOLD.

Rather than repeat the interchange instructions, however, a more elegant approach is to use a separate subroutine to perform the interchange. Such a subroutine was included in Figure 4.4. We will examine this approach in our second version of the bubble sort routine.

FORTRAN PROGRAM: BUBBLE SORT WITH SWAP

A revised main program for testing the sorting routines is given in Figure 6.4. This program can be used to test all the sorting routines given in this chapter. The new version will not execute any faster than the previous one, but the source code is easier to comprehend.

A new version of the bubble sort is given in Figure 6.5. This version will run much faster than the first one if the original array is already ordered. Otherwise, it will run more slowly. Another feature of this bubble sort is a call to subroutine SWAP for the interchange of two elements. With this change, the sorting routine can easily be altered so that additional arrays such as B and C can be easily sorted along with array A. The new lines:

```
CALL SWAP(B(I), B(J))
CALL SWAP(C(I), C(J))
```

can be added immediately after the first call to subroutine SWAP for vector A.

```
C         PROGRAM BSORT2
C
C -- Test sort routine.
C -- Function RAN needed.
C -- May  2, 81
C
        INTEGER LENGTH, IN, OUT, MAX, I
        REAL X(500)
        COMMON /INOUT/ IN, OUT, MAX
C
        IN = 1
        OUT = 1
        MAX = 500
10      CALL INPUT(X, LENGTH)
        WRITE(OUT, 105) (X(I), I = 1, LENGTH)
C -- Random list
        CALL OUTPUT(X, LENGTH)
        WRITE(OUT, 102)
C -- Sorted list
        CALL OUTPUT(X, LENGTH)
        WRITE(OUT, 103)
        DO 20 I = 1, LENGTH
          X(I) = LENGTH + 1 - I
20      CONTINUE
C -- Reversed list
        CALL OUTPUT(X, LENGTH)
        WRITE(OUT, 104)
        GOTO 10
101     FORMAT(1X, A4)
102     FORMAT(' Random')
103     FORMAT(' Sorted')
104     FORMAT(' Reversed')
105     FORMAT(1X, 10F7.2)
        END
        SUBROUTINE INPUT(X, N)
C
C -- Get N and generate random vector X.
C
        INTEGER N, IN, OUT, I, J, MAX
        REAL X(1)
        COMMON /INOUT/ IN, OUT, MAX
C
5       WRITE(OUT, 101)
        READ(IN,102) N
        IF (N .GT. MAX) GOTO 5
        IF (N .LT. 0) GOTO 99
        DO 10 I = 1, N
          X(I) = RAN(0) * 100.0
10      CONTINUE
        RETURN
99      STOP
101     FORMAT(' How many points? ')
102     FORMAT(I3)
```

Figure 6.4: An Improved Driver for the Sorting Routine

```
          END
          SUBROUTINE OUTPUT(X, N)
C
C -- Print the vector X.
C
          INTEGER N, IN, OUT, I, BELL, BEL
          REAL X(1)
          COMMON /INOUT/ IN, OUT
          DATA BELL/'G'/
C
          BELL = BELL - 64
          WRITE(OUT, 101) BELL
          CALL SORT(X, N)
          WRITE(OUT, 101) BELL
          WRITE(OUT, 102) (X(I), I = 1, N)
          RETURN
101       FORMAT(1X, A4)
102       FORMAT(1X, 10F7.2)
          END
```

Figure 6.4: An Improved Driver for the Sorting Routine (cont.)

```
          SUBROUTINE SORT(A, N)
C
C -- A variation of bubble sort.
C -- May  2, 81
C
          LOGICAL CHANGE
          INTEGER N, J, N1
          REAL A(1)
C
          N1 = N - 1
10        CHANGE = .FALSE.
             DO 20 J = 1, N1
                IF (A(J) .LE. A(J+1)) GOTO 20
                   CALL SWAP(A(J), A(J+1))
                   CHANGE = .TRUE.
20           CONTINUE
          IF (CHANGE) GOTO 10
          RETURN
          END
```

Figure 6.5: A Variation of the Bubble Sort, Incorporating Function SWAP

Running the New Version

Make a copy of the driver program given in Figure 6.4 and a separate copy of the new bubble sort given in Figure 6.5. Separately compile the two parts and then execute them. Compare the sorting times to the first version.

Next we will study a more sophisticated and slightly more difficult sorting routine.

A SHELL SORT

The major disadvantage with the bubble sort method is that it frequently makes more comparisons and more interchanges than are necessary. An item may be moved from one end of the array to the other. Then, a little later, it might be moved nearly back to where it started.

The *Shell-Metzner* sort (or *Shell* sort) is generally more efficient than the bubble sort. Comparisons are initially made over long distances. The first item in the array is compared to one in the middle, rather than to the one right next door. For short lists of fewer than a dozen items, the Shell sort and bubble sort are comparable in speed. But as the length of the list increases, the speed of the Shell sort becomes apparent. The Shell sort is noticeably faster than the bubble sort when the number of items exceeds about fifty.

We will be able to quantify our comparison of the bubble and shell sorts by comparing the sorting times.

FORTRAN PROGRAM: THE SHELL-METZNER SORT

Make a copy of the Shell sort subroutine given in Figure 6.6. Use a file name that is different from the two previous sorting routines. You can use the previous driver with the new Shell sort without recompiling the driver. Run the new version and compare the sorting times to the bubble sort

```
           SUBROUTINE SORT(A, N)
C
C -- Shell-Metzner sort for vector A
C -- May 19, 81
C
           INTEGER N, I, J, JUMP, J2, J3
           REAL A(1)
C
           JUMP = N
10         JUMP = JUMP / 2
             IF (JUMP .EQ. 0) GOTO 99
               J2 = N - JUMP
             DO 30 J = 1, J2
15           I = J
20           J3 = I + JUMP
               IF (A(I) .LE. A(J3)) GOTO 30
               CALL SWAP(A(I), A(J3))
               I = I - JUMP
               IF (I .GT. 0) GOTO 20
30         CONTINUE
           GOTO 10
99         RETURN
           END
```

Figure 6.6: A Shell Sort Subroutine

routines. You should find that the Shell sort runs much faster. You will also notice that the Shell sort, like the bubble sort, runs much faster with sorted data than with unsorted data.

In the discussion of our last and most complex sorting routine we will take up the issue of *recursion,* and we will present a nonrecursive version of the quick sort algorithm.

THE QUICK SORT

We saw that the bubble sort is easy to program and easy to understand. The Shell sort is a bit more complicated, but it can sort much more quickly. Both of these two algorithms can readily be programmed in other high-level languages such as Pascal or BASIC. The third sorting algorithm we will consider is known as the *quick* sort. It is even more complicated than the previous algorithms. It is generally faster than the bubble sort or the Shell sort, although if the original data are already sorted, or nearly sorted, the Shell sort can be much faster. A quick sort routine takes almost as long to run on a sorted array as on an unsorted one. It may take considerably longer with a list arranged in reverse order.

FORTRAN PROGRAM: A NONRECURSIVE QUICK SORT

The bubble sort begins by comparing elements that are side by side. The Shell sort begins by comparing the initial element to one at the middle of the array. The quick sort begins by comparing the elements at the opposite ends of the array. Thus, the initial interchanges for the quick sort can be made over long distances.

An array to be sorted is repeatedly partitioned into two smaller and smaller parts. The elements are rearranged so that all the elements on one side are smaller than all those on the other. The two new sections are then divided into two subsections and the process is repeated. Partitioning continues until there are many sets containing one element each, at which point the array is sorted.

The quick sort algorithm is usually written to operate *recursively;* that is, one portion of the code calls itself. Recursion can be readily implemented in certain computer languages. However, recursive subroutines are not allowed in FORTRAN. For this reason, the version of the quick sort given in Figure 6.7 is nonrecursive. This algorithm simulates the recursive calls by saving the partition pointers in a stack. The left and right pointers for the simulated calls are stored in the integer arrays LEFT and RIGHT.

The pivot element is initially chosen to be the last element in the array. This is a poor choice, however, if the array is already arranged in increasing or decreasing order. Consequently, this pivot element is compared with two other elements: the first element of the array and the element located at the center of the array. The element that is the median of these three, neither the largest nor the smallest, is chosen as the pivot. This change will greatly speed up the sorting process when the array or subarray is already ordered in one direction or the other.

One feature of this version is that at each partitioning, the smaller of the two subsets is sorted before the larger. This minimizes the space needed for storage of the unsorted indices.

You will want to keep versions of both the Shell sort and the quick sort and use the one you find to be fastest.

```
          SUBROUTINE SORT(A, N)
C
C -- Nonrecursive quick sort for vector A.
C -- May  3, 81
C
          INTEGER N, I, J, SP, MID, LEFT(20), RIGHT(20)
          REAL A(1), PIVOT
C
          LEFT(1) = 1
          RIGHT(1) = N
          SP = 1
10        IF (LEFT(SP) .LT. RIGHT(SP)) GOTO 20
              SP = SP - 1
              GOTO 130
20        I = LEFT(SP)
          J = RIGHT(SP)
          PIVOT = A(J)
          MID = (I + J) / 2
          IF (J - I .LT. 6) GOTO 50
          IF ((PIVOT .GT. A(I)).AND.(PIVOT .LT. A(MID))) GOTO 50
          IF ((PIVOT .LT. A(I)).AND.(PIVOT .GT. A(MID))) GOTO 50
          IF ((A(I) .LT. A(MID)).AND.(A(I) .GT. PIVOT)) GOTO 30
          IF ((A(I) .GT. A(MID)).AND.(A(I) .LT. PIVOT)) GOTO 30
            CALL SWAP(A(MID), A(J))
            GOTO 40
30          CALL SWAP(A(I), A(J))
40        PIVOT = A(J)
50        IF (I .GE. J) GOTO 110
60          IF (A(I) .GE. PIVOT) GOTO 70
              I = I + 1
              GOTO 60
70          J = J - 1
80          IF (.NOT.((I .LT. J) .AND. (PIVOT .LT. A(J)))) GOTO 90
              J = J - 1
              GOTO 80
90          IF (I .LT. J) CALL SWAP(A(I), A(J))
100         GOTO 50
110       J = RIGHT(SP)
          CALL SWAP(A(I), A(J))
```

Figure 6.7: A Nonrecursive Quick Sort

```
         IF ((I - LEFT(SP)) .GE. (RIGHT(SP) - I)) GOTO 120
C -- Stack shorter first.
         LEFT(SP+1) = LEFT(SP)
         RIGHT(SP+1) = I - 1
         LEFT(SP) = I + 1
         GOTO 125
120      LEFT(SP+1) = I + 1
         RIGHT(SP+1) = RIGHT(SP)
         RIGHT(SP) = I - 1
125   .  SP = SP + 1
130      IF (SP .GT. 0) GOTO 10
         RETURN
         END
```

Figure 6.7: A Nonrecursive Quick Sort (cont.)

Finally, let us return to our curve-fitting program, which originally led us into this discussion of sorting routines.

INCORPORATING SORT INTO THE CURVE-FITTING PROGRAM

The next step is to incorporate one of the sorting routines into the curve-fitting program we wrote in the previous chapter. As before, keep a working copy of the prior version. Then you will have something to go back to if the new version becomes hopelessly mixed up. Two changes will have to be made to the sorting routine so that the array Y is sorted along with the array X. One change is to add a second array name to the parameter list at the beginning of the program (we will use the name B in our example). The second change adds another call to subroutine SWAP so that an interchange of two elements of vector A will also interchange the corresponding elements of vector B. The heading becomes:

SUBROUTINE SORT(A, B, N)

If you have chosen the Shell sort, then immediately after the statement:

CALL SWAP(A(I), A(J3))

place the statement:

CALL SWAP(B(I), B(J3))

A call to the sorting subroutine is inserted just after the call to the input subroutine at the beginning of the main program:

CALL SORT(X, Y, LENGTH)

If, on the other hand, a call to the sorting routine is delayed until after the linear fit is calculated, then we will have to sort three arrays: X, Y, and YCALC.

SUMMARY

We have seen variations of three common sort routines: the bubble sort, the Shell sort and the quick sort. All of these sorting routines are written to operate on real numbers. Keep in mind that these routines can be easily altered to sort integers, logical variables, or strings of characters.

Now that our curve-fitting program can handle real experimental data, we are ready to apply the program to some more complex equations. We will do this in Chapter 7.

EXERCISES

6-1: *Incorporate into one of the sorting routines a direction flag, as a logical variable called UPFLAG. Preset the value of UPFLAG to .TRUE. at the beginning of the sort routine. Change the comparison step so that sorting will occur in the usual way if UPFLAG is true, but sort the data in descending order if the flag is false. For example, there might be a pair of expressions such as:*

IF (UPFLAG .AND. (X(I) .LE. A(J3))) GOTO 30

IF (.NOT. UPFLAG .AND. (X(I) .GE. A(J3))) GOTO 30

Use the sign of the last parameter, the number of items N, to signal a reverse sort. If N is negative, make UPFLAG false and restore the sign of N to positive. The main program should ask the user whether sorting is to be performed in ascending or descending order. If the items are to be sorted in reverse order, make N negative in the calling parameter list.

7

General Least-Squares Curve Fitting

IN THIS CHAPTER we will develop several different least-squares curve-fitting programs. Our goal will be to *generalize* the curve-fitting program developed in Chapter 5 so that it will be a more useful and realistic tool for a wider variety of experimental situations. Up to now we have been limited to the equation of a straight line. Our first program in this chapter will implement a parabolic curve—that is, a second-order polynomial equation. From that point we will develop a method that will handle both higher-order polynomials and nonpolynomial equations. The only restriction will be that the unknown coefficients must be linear.

The key to our approach will be the use of a data vector and a data matrix. This method will subsequently allow us to input the *order* of a polynomial equation from the keyboard once the program is running. Finally, we will implement curve-fitting programs for some real experimental data.

The equations of the fitted data will include those used for heat capacity and vapor pressure. In addition, in our program for a three-variable equation of state, we will experiment with solutions for a nonlinear coefficient.

A PARABOLIC CURVE FIT

A least-squares curve-fitting program was developed in Chapter 5. That program was used to calculate the coefficients A and B for the expression:

$$y = A + Bx \tag{1}$$

While Equation 1 is the most commonly used curve-fitting equation, there are times when a different equation is required. Therefore, in this chapter, we will extend the least-squares method to include other commonly used expressions.

We will place one restriction on the form of the curve-fitting equation. It must be linear in the unknown coefficients. Thus, we can consider equations such as:

$$y = A + Bx + Cx^2$$

$$y = A + \frac{B}{x} + Cz$$

and

$$\ln y = \frac{Ax}{z} + B\,e^x$$

because the coefficients A, B, and C are linear. But we will not consider an equation such as:

$$y = A + B\,e^{Cx}$$

since the coefficient C is not linear.

The development of a curve-fitting program for any of the above equations is similar to the approach we used in Chapter 5. Consider, for example, the parabolic equation which is a second-order polynomial:

$$y = A + Bx + Cx^2 \tag{2}$$

We define the residuals to be:

$$r = A + Bx + Cx^2 - y$$

The residuals are squared, and then summed. As in Chapter 5, we wish to minimize this quantity. We thus take the derivative with respect to each variable (A, B, and C) and set the resulting equations to zero. For the parabolic curve fit there will be three equations, one for each of the three variables. The resulting equations are:

$$An + B\Sigma x + C\Sigma x^2 = \Sigma y \tag{3}$$

$$A\Sigma x + B\Sigma x^2 + C\Sigma x^3 = \Sigma xy \tag{4}$$

$$A\Sigma x^2 + B\Sigma x^3 + C\Sigma x^4 = \Sigma x^2 y \tag{5}$$

FORTRAN PROGRAM: LEAST-SQUARES CURVE FIT FOR A PARABOLA

The solution to the parabolic fit is obtained by solving these three equations simultaneously. The program shown in Figure 7.1 finds the solution to these equations by using Cramer's rule. The determinants of four 3-by-3 matrices must be solved with this approach. One function subprogram and three subroutines we developed in previous chapters are used by this program. However, these routines are not included in the listing. The necessary routines are DETERM (Figure 3.3), CRAMER and SETUP (Figure 4.3), and PLOT (Figure 5.3).

In the previous chapters, the coefficients were represented by the variables A, B, and C. However, in this chapter, we will use the vector COEF for this purpose. Thus, the constant term, A, now corresponds to the first element of the vector, COEF(1). Similarly, the coefficients B and C correspond to COEF(2) and COEF(3), respectively.

```
C       PROGRAM LEAST1
C
C  --  Parabolic least-squares fit.
C  --  Function DETERM and
C  --  subroutines CRAMER, SETUP, and PLOT required.
C  --  May 3, 81
C
        INTEGER IN, OUT, MAXR, MAXC, LINES, NROW, NCOL
        REAL X(20), Y(20), YCALC(20), COEF(3), CORREL
        COMMON /INOUT/ IN, OUT
C
        IN = 1
        OUT = 1
        MAXR = 20
        MAXC = 3
        CALL INPUT(X, Y, NROW)
        CALL LINFIT(X, Y, YCALC, COEF, NROW, NCOL, CORREL)
        CALL OUTPUT(X, Y, YCALC, COEF, NROW, NCOL, CORREL)
        LINES = 2*(NROW - 1) + 1
        CALL PLOT(X, Y, YCALC, NROW, OUT, LINES)
        STOP
        END
        SUBROUTINE INPUT(X, Y, NROW)
C  --  Get values for NROW and arrays X and Y.
C
        INTEGER NROW, I
        REAL X(1), Y(1)
C
        NROW = 9
        DO 10 I = 1, NROW
          X(I) = I
10      CONTINUE
```

Figure 7.1: A Parabolic Least-Squares Fit

```
          Y(1) =   2.07
          Y(2) =   8.6
          Y(3) =  14.42
          Y(4) =  15.8
          Y(5) =  18.92
          Y(6) =  17.96
          Y(7) =  12.98
          Y(8) =   6.45
          Y(9) =   0.27
          RETURN
          END
          SUBROUTINE OUTPUT(X, Y, YCALC, COEF, NROW, NCOL, CORREL)
C
C -- Print out the answers.
C
          INTEGER IN, OUT, NROW, NCOL, I
          REAL X(1), Y(1), YCALC(1), COEF(1), CORREL
          COMMON /INOUT/ IN, OUT
C
          WRITE(OUT, 101)
          WRITE(OUT, 102) (I, X(I), Y(I), YCALC(I), I = 1, NROW)
          WRITE(OUT, 103)
          WRITE(OUT, 104) (COEF(I), I = 1, NCOL)
          WRITE(OUT, 105) CORREL
          RETURN
101       FORMAT('   I        X         Y       Y CALC')
102       FORMAT(I4, F8.1, 2F9.2)
103       FORMAT(/' Coefficients')
104       FORMAT(F9.4)
105       FORMAT(/' Correlation coefficient is', F7.4)
          END
          SUBROUTINE LINFIT(X, Y, YCALC, COEF, NROW, NCOL, COR)
C
C -- Least squares fit to a parabola.
C -- NROW pairs of X-Y points.
C
          LOGICAL ERROR
          INTEGER NROW, NCOL, I
          REAL X(1), Y(1), YCALC(1), COEF(1), SXY, XYY, SXX
          REAL A(3,3), G(3)
          REAL SUMX, SUMY, SUMXY, SUMX2, SUMY2, XI, YI
          REAL SUMX3, SUMX4, DENOM, SRS, X2, RES, COR
C
          NCOL = 3
          SUMX = 0.0
          SUMY = 0.0
          SUMXY = 0.0
          SUMX2 = 0.0
          SUMY2 = 0.0
          SUMX3 = 0.0
          SUMX4 = 0.0
          SUM2Y = 0.0
```

Figure 7.1: A Parabolic Least-Squares Fit (cont.)

```
        DO 10 I = 1, NROW
          XI = X(I)
          YI = Y(I)
          X2 = XI * XI
          SUMX = SUMX + XI
          SUMY = SUMY + YI
          SUMXY = SUMXY + XI * YI
          SUMX2 = SUMX2 + X2
          SUMY2 = SUMY2 + YI * YI
          SUMX3 = SUMX3 + XI * X2
          SUMX4 = SUMX4 + X2 * X2
          SUM2Y = SUM2Y + X2 * YI
10      CONTINUE
        A(1,1) = NROW
        A(2,1) = SUMX
        A(1,2) = SUMX
        A(3,1) = SUMX2
        A(1,3) = SUMX2
        A(2,2) = SUMX2
        A(3,2) = SUMX3
        A(2,3) = SUMX3
        A(3,3) = SUMX4
        G(1) = SUMY
        G(2) = SUMXY
        G(3) = SUM2Y
        CALL CRAMER(A, G, COEF, ERROR)
        SRS = 0.0
        DO 20 I = 1, NROW
          YCALC(I) = COEF(1) + COEF(2)*X(I) + COEF(3)*X(I)*X(I)
          RES = Y(I) - YCALC(I)
          SRS = SRS + RES*RES
20      CONTINUE
        COR = SQRT(1.0 - SRS/(SUMY2 - SUMY*SUMY/NROW))
        RETURN
        END
```

Figure 7.1: A Parabolic Least-Squares Fit (cont.)

The plotting subroutine requires the array of independent variables to be arranged in increasing or decreasing order. The data generated by the input subroutine are arranged in decreasing order. However, at a later time, you may want to substitute other data that are not sorted. Then you can include one of the sorting subroutines developed in Chapter 6.

Running the Program

Type up this program and run it. The results should look like Figure 7.2. The correlation coefficient is close to unity, indicating that the parabolic equation can produce a good fit to the data. The resulting equation is:

$$y = -7.827 + 10.59x - 1.083x^2$$

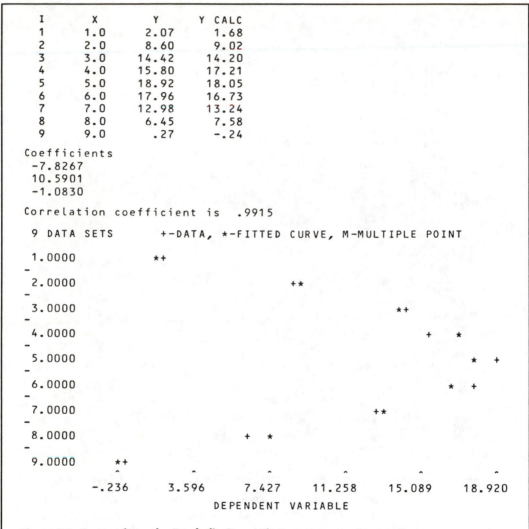

Figure 7.2: Output from the Parabolic Curve Fit Program

Now let us consider polynomial equations with orders higher than 2, and nonpolynomial equations. We will see that the approach we have been using to solve for the coefficients becomes less practical as the equation becomes more complex. Therefore, we will want to investigate another method.

CURVE FITS FOR OTHER EQUATIONS

If a higher-order polynomial is chosen, there will be additional coefficients, resulting in additional equations to be solved simultaneously.

Then, Equations 3, 4, and 5 can be easily extended. For example, to find the coefficients to the cubic equation:

$$y = A + Bx + Cx^2 + Dx^3$$

the residuals are defined as:

$$r = A + Bx + Cx^2 + Dx^3 - y$$

The residuals are first squared and then summed. The derivative is taken with respect to each of the four variables A, B, C, and D. The resulting four equations are solved simultaneously:

$$An + B\Sigma x + C\Sigma x^2 + D\Sigma x^3 = \Sigma y$$
$$A\Sigma x + B\Sigma x^2 + C\Sigma x^3 + D\Sigma x^4 = \Sigma xy$$
$$A\Sigma x^2 + B\Sigma x^3 + C\Sigma x^4 + D\Sigma x^5 = \Sigma x^2 y$$
$$A\Sigma x^3 + B\Sigma x^4 + C\Sigma x^5 + D\Sigma x^6 = \Sigma x^3 y$$

Nonpolynomial expressions are treated similarly. For example, the general three-term equation is:

$$y = A + Bf(x) + Cg(x)$$

where $f(x)$ and $g(x)$ represent any function of x. The residuals, defined as:

$$r = A + B f(x) + C g(x) - y$$

are squared and then summed. The derivatives with respect to the three variables give the three equations:

$$An + B\Sigma f(x) + C\Sigma g(x) = \Sigma y$$
$$A\Sigma f(x) + B\Sigma f(x)^2 + C\Sigma f(x)g(x) = \Sigma f(x)y$$
$$A\Sigma g(x) + B\Sigma f(x)g(x) + C\Sigma g(x)^2 = \Sigma g(x)y$$

While the above approach to curve fitting is correct, it is laborious. Major alterations are needed whenever the curve-fitting equation is changed. For example, if the equation were chosen to be:

$$y = A + Bx + \frac{C}{x^2}$$

then there would need to be statements in the program such as:

SUMX5 = SUMX5 + 1 / XI

and

SUMX6 = SUMX6 + 1 / (XI * XI)

for calculating the needed sums.

A Direct Solution

A better way to determine the coefficients of the curve-fitting equation is to set up the data in a matrix and a vector. The data matrix and data vector are then converted to a set of simultaneous equations that are solved by methods we developed in Chapter 4. For example, suppose that we want a linear fit to the equation:

$$y = A + Bx$$

for five sets of x-y data. In this example, the data vector would simply be the vector of y values. The data matrix would look like this:

$$\begin{bmatrix} 1 & x_1 \\ 1 & x_2 \\ 1 & x_3 \\ 1 & x_4 \\ 1 & x_5 \end{bmatrix}$$

Each row of the data matrix corresponds to one data point, whereas each column of the matrix corresponds to one term of the equation. Consequently, the data matrix has five rows and two columns. Column 1 of the data matrix contains only the value of 1, since that is the corresponding function of x (i.e., x^0) in the first term of the equation. Column 2 contains the values of x, because that is the function of x in the second term of the equation.

The data matrix for a parabolic fit would have three columns. The first two columns would be the same as they were in the straight-line fit. The third column, however, would contain the square of each x value. If, on the other hand, we chose an equation like:

$$\ln p = A + \frac{B}{t} + C \ln t$$

then the first column of the data matrix would contain the value of 1, the second column would have the reciprocal of the data, and the third column would contain the logarithm of the data. The data vector in this case would have the logarithm of p.

The rectangular data matrix is converted to a square matrix by using a simple operation. The transpose of the data matrix is multiplied by the matrix itself to produce the coefficient matrix. For the straight-line fit of five sets of x-y data, the operation is:

$$\begin{bmatrix} 1 & 1 & 1 & 1 & 1 \\ x_1 & x_2 & x_3 & x_4 & x_5 \end{bmatrix} \begin{bmatrix} 1 & x_1 \\ 1 & x_2 \\ 1 & x_3 \\ 1 & x_4 \\ 1 & x_5 \end{bmatrix}$$

The result is a 2-by-2 matrix containing the required sums of x:

$$\begin{bmatrix} n & \Sigma x \\ \Sigma x & \Sigma x^2 \end{bmatrix}$$

The product of the data vector (considered as a row vector) and the data matrix:

$$\begin{bmatrix} y_1 & y_2 & y_3 & y_4 & y_5 \end{bmatrix} \begin{bmatrix} 1 & x_1 \\ 1 & x_2 \\ 1 & x_3 \\ 1 & x_4 \\ 1 & x_5 \end{bmatrix}$$

gives the required constant vector of length 2:

$$\begin{bmatrix} \Sigma y & \Sigma xy \end{bmatrix}$$

We will now examine a FORTRAN implementation of this approach. We will also want to incorporate into the program a general technique for calculating the standard error. As we predicted in Chapter 4, the Gauss-Jordan method of solving simultaneous equations will prove to be an important tool to use here because it supplies the *inverse* of the coefficient matrix.

FORTRAN PROGRAM: THE MATRIX APPROACH TO CURVE FITTING

Figure 7.3 gives a curve-fitting program that utilizes this matrix approach. The data matrix and data vector are set up in subroutine LINFIT. Then subroutine SQUARE is used to convert the data matrix, X, and data vector, $\mathbf{y}$, into the square coefficient matrix, A, and the constant vector, $\mathbf{g}$. The operations are:

$$X^T X = A$$

and

$$\mathbf{y}X = \mathbf{g}$$

The data vector is multiplied by the data matrix to produce the constant vector. The solution vector:

$$A^{-1}\mathbf{g} = \mathbf{b}$$

can be obtained by using any of the routines we developed in Chapter 4 for the solution of simultaneous equations.

The linear curve-fitting program developed in Chapter 5 presented the standard errors along with the corresponding elements of the solution to the approximating function. The standard errors were obtained from the standard error of the estimate and the summation of x and x^2. We will use a more general technique at this point.

The standard errors are readily obtained from the inverse of the co-efficient matrix, A. The value corresponding to the ith term of the approximating function is the product of the standard error of the estimate and the square root of the ith term of the major diagonal of the inverse:

$$\begin{bmatrix} a_{11} & & \\ & a_{22} & \\ & & a_{33} \end{bmatrix}^{-1}$$

The FORTRAN expression is

$$\text{SIGMA(I)} = \text{SEE} * \text{SQRT(A(I,I))}$$

Of the several methods we previously considered for the simultaneous solution of linear equations, only the Gauss-Jordan method generated the inverse of the coefficient matrix. Since we need this inverse to determine the errors on the elements of the solution vector, the Gauss-Jordan method is the natural choice. Consequently, we will use this method for the remaining curve-fitting programs in this chapter.

The standard errors on the coefficients can be used to determine the confidence intervals for the corresponding coefficients. In addition, the standard errors can alert us to the possibility of ill conditioning. We discussed ill-conditioned matrices in Chapter 4. They are more likely to be encountered during the solution of simultaneous equations than in curve fitting. Nevertheless, we will want to watch for such a problem. The standard errors are derived from the square roots of the diagonal elements of the inverted matrix. Large differences in these elements would suggest ill conditioning. Therefore, if the squares of the errors are many orders apart, then ill conditioning may be present. The last program in this chapter demonstrates ill conditioning.

The program shown in Figure 7.3 is similar to the one given in Figure 7.1, but the solution to the curve-fitting equation is determined by the more general matrix method. Three previously developed routines are used: the matrix multiplication subroutine (SQUARE), given in Chapter 3; the Gauss-Jordan subroutine (GAUSSJ), developed in Chapter 4; and the plotting routine (PLOT), from Chapter 5.

```
C       PROGRAM LEAST2
C
C -- Parabolic least-squares fit.
C -- Subroutines GAUSSJ, SQUARE, and PLOT required.
C -- May 3, 81
```

Figure 7.3: A Parabolic Least-Squares Fit Using Gauss-Jordan Elimination

```
C
        INTEGER IN, OUT, MAXR, MAXC, LINES, NROW, NCOL
        REAL X(20), Y(20), YCALC(20), COEF(3), CORREL
        REAL SIG(3), RESID(20)
        COMMON /INOUT/ IN, OUT, MAXR, MAXC
C
        IN = 1
        OUT = 1
        MAXR = 20
        MAXC = 3
        CALL INPUT(X, Y, NROW)
        CALL LINFIT(X, Y, YCALC, RESID, COEF, SIG,
     *    NROW, NCOL, CORREL)
        CALL OUTPUT(X, Y, YCALC, RESID, COEF, SIG,
     *    NROW, NCOL, CORREL)
        LINES = 2*(NROW - 1) + 1
        CALL PLOT(X, Y, YCALC, NROW, OUT, LINES)
        STOP
        END
        SUBROUTINE INPUT(X, Y, NROW)
C -- Get values for NROW and arrays X and Y.
C
        INTEGER NROW, I
        REAL X(1), Y(1)
C
        NROW = 9
        DO 10 I = 1, NROW
          X(I) = I
10      CONTINUE
        Y(1) =   2.07
        Y(2) =   8.6
        Y(3) =  14.42
        Y(4) =  15.8
        Y(5) =  18.92
        Y(6) =  17.96
        Y(7) =  12.98
        Y(8) =   6.45
        Y(9) =   0.27
        RETURN
        END
        SUBROUTINE OUTPUT(X, Y, YCALC, RESID, COEF, SIG,
     *    NROW, NCOL, CORREL)
C
C -- Print out the answers.
C
        INTEGER IN, OUT, NROW, NCOL, I
        REAL X(1), Y(1), YCALC(1), COEF(1), CORREL
        REAL RESID(1), SIG(1)
        COMMON /INOUT/ IN, OUT
C
        WRITE(OUT, 101)
        WRITE(OUT, 102) (I, X(I), Y(I), YCALC(I),
```

Figure 7.3:
A Parabolic Least-Squares Fit Using Gauss-Jordan Elimination (cont.)

```
      *  RESID(I), I = 1, NROW)
         WRITE(OUT, 103)
         WRITE(OUT, 106) COEF(1), SIG(1)
         WRITE(OUT, 104) (COEF(I), SIG(I), I = 2, NCOL)
         WRITE(OUT, 105) CORREL
         RETURN
101      FORMAT(' I        X         Y       Y CALC')
102      FORMAT(I4, F8.1, 3F9.2)
103      FORMAT(/' Coefficients     Errors')
104      FORMAT(0PF8.3, 3X, 1PE12.3)
105      FORMAT(/' Correlation coefficient is', F7.4)
106      FORMAT(0PF8.3, 3X, 1PE12.3, '  Constant term')
         END
         SUBROUTINE LINFIT(X, Y, YCALC, RESID, COEF, SIG,
      *  NROW, NCOL, COR)
C
C -- Least squares fit to NROW sets of X-Y points
C -- using Gauss-Jordan elimination.
C -- Subroutines SQUARE and GAUSSJ needed.
C
         LOGICAL ERROR
         INTEGER NROW, NCOL, I, J, MAXR, MAXC, IN, OUT
         INTEGER INDEX(20,3), NVEC
         REAL X(1), Y(1), YCALC(1), COEF(1), RESID(1)
         REAL A(3,3), XMATR(20,3), SIG(1)
         REAL SUMY, SUMY2, XI, YI, YC, RES, COR, SRS, SEE
         COMMON /INOUT/ IN, OUT, MAXR, MAXC
         DATA NVEC/1/
C
         NCOL = 3
         DO 10 I = 1, NROW
            XI = X(I)
            XMATR(I,1) = 1.0
            XMATR(I,2) = XI
            XMATR(I,3) = XI * XI
10       CONTINUE
         CALL SQUARE(XMATR, Y, A, COEF, NROW, NCOL, MAXR, MAXC)
         CALL GAUSSJ(A, COEF, INDEX, NCOL, MAXC, NVEC, ERROR, OUT)
         SUMY = 0.0
         SUMY2 = 0.0
         SRS = 0.0
         DO 20 I = 1, NROW
            YI = Y(I)
            YC = 0.0
            DO 15 J = 1, NCOL
15             YC = YC + COEF(J) * XMATR(I,J)
            YCALC(I) = YC
            RES = YC - YI
            RESID(I) = RES
            SRS = SRS + RES*RES
            SUMY = SUMY + YI
            SUMY2 = SUMY2 + YI * YI
20       CONTINUE
```

Figure 7.3: A Parabolic Least-Squares Fit Using Gauss-Jordan Elimination (cont.)

```
         COR = SQRT(1.0 - SRS/(SUMY2 - SUMY*SUMY/NROW))
         IF (NROW .EQ. NCOL) SEE = SQRT(SRS)
         IF (NROW .NE. NCOL) SEE = SQRT(SRS / (NROW - NCOL))
         DO 30 I = 1, NCOL
   30       SIG(I) = SEE * SQRT(A(I,I))
         RETURN
         END
```

Figure 7.3: A Parabolic Least-Squares Fit Using Gauss-Jordan Elimination (cont.)

Running the Program

Alter the first program in this chapter or create a new one so that it looks like the one given in the listing of Figure 7.3. Be sure to have available the Gauss-Jordan and plotting routines and subroutine SQUARE. Run the program and compare the output with Figure 7.4. The results should be the same as they were for the previous program, except that the residuals and errors are shown for the new version.

```
        I        X         Y        Y CALC      RESID
        1       1.0      2.07       1.68        -.39
        2       2.0      8.60       9.02         .42
        3       3.0     14.42      14.20        -.22
        4       4.0     15.80      17.21        1.41
        5       5.0     18.92      18.05        -.87
        6       6.0     17.96      16.73       -1.23
        7       7.0     12.98      13.24         .26
        8       8.0      6.45       7.58        1.13
        9       9.0       .27       -.24        -.51

    Coefficients      Errors
      -7.827         1.298E+00    Constant term
      10.590         5.960E-01
      -1.083         5.813E-02

    Correlation coefficient is   .9915
```

Figure 7.4:
Output: An Alternate Version of a Parabolic Least- Squares Curve Fit

In the next section the matrix approach will permit us to try out different orders of polynomial equations to fit any given set of data. To develop a sense of the full power of this tool, we will run this new program several times on one set of data. By comparing the resulting plotted curves and correlation coefficients, we will find the best polynomial order to fit our data.

FORTRAN PROGRAM: ADJUSTING THE ORDER OF THE POLYNOMIAL

One advantage of the new version of our curve-fitting program is that it is easy to change both the number of rows, corresponding to the number of data points, and the number of columns, which corresponds to the number of polynomial terms in the curve-fitting equation. Consequently, for this third version, we will proceed one step further in "generalizing" our program. We will input the order of the polynomial equation from the console. The order, of course, is one smaller than the number of terms in the equation. Make a copy of the previous program and alter it so that it looks like the one shown in Figure 7.5.

```
C       PROGRAM LEAST3
C
C -- Parabolic least-squares fit.
C -- Input polynomial order from console.
C -- Subroutines GAUSSJ, SQUARE, and PLOT required.
C -- May 3, 81
C
        INTEGER IN, OUT, MAXR, MAXC, LINES, NROW, NCOL
        REAL X(20), Y(20), YCALC(20), COEF(3), CORREL
        REAL SIG(3), RESID(20)
        COMMON /INOUT/ IN, OUT, MAXR, MAXC
C
        IN = 1
        OUT = 1
        MAXR = 20
        MAXC = 3
10      CALL INPUT(X, Y, NROW, NCOL)
        CALL LINFIT(X, Y, YCALC, RESID, COEF, SIG,
     *   NROW, NCOL, CORREL)
        CALL OUTPUT(X, Y, YCALC, RESID, COEF, SIG,
     *   NROW, NCOL, CORREL)
        LINES = 2*(NROW - 1) + 1
        CALL PLOT(X, Y, YCALC, NROW, OUT, LINES)
        GOTO 10
        END
        SUBROUTINE INPUT(X, Y, NROW, NCOL)
C -- Get values for NROW and arrays X and Y.
C
        INTEGER NROW, I, IN, OUT, MAXR, MAXC, NCOL
        REAL X(1), Y(1)
        COMMON /INOUT/ IN, OUT, MAXR, MAXC
C
5       WRITE(OUT, 101)
        READ(IN, 102) NCOL
        IF (NCOL .GT. MAXC - 1) GOTO 5
        IF (NCOL .LT. 1) GOTO 99
```

Figure 7.5: Input Polynomial Order from the Console

```
        NCOL = NCOL + 1
        NROW = 9
        DO 10 I = 1, NROW
           X(I) = I
10      CONTINUE
        Y(1) =   2.07
        Y(2) =   8.6
        Y(3) =  14.42
        Y(4) =  15.8
        Y(5) =  18.92
        Y(6) =  17.96
        Y(7) =  12.98
        Y(8) =   6.45
        Y(9) =   0.27
        RETURN
99      STOP
101     FORMAT(' Order? ')
102     FORMAT(I2)
        END
        SUBROUTINE OUTPUT(X, Y, YCALC, RESID, COEF, SIG,
     *  NROW, NCOL, CORREL)
C
C -- Print out the answers.
C
        INTEGER IN, OUT, NROW, NCOL, I
        REAL X(1), Y(1), YCALC(1), COEF(1), CORREL
        REAL RESID(1), SIG(1)
        COMMON /INOUT/ IN, OUT
C
        WRITE(OUT, 101)
        WRITE(OUT, 102) (I, X(I), Y(I), YCALC(I),
     *  RESID(I), I = 1, NROW)
        WRITE(OUT, 103)
        WRITE(OUT, 106) COEF(1), SIG(1)
        WRITE(OUT, 104) (COEF(I), SIG(I), I = 2, NCOL)
        WRITE(OUT, 105) CORREL
        RETURN
101     FORMAT('   I      X       Y      Y CALC')
102     FORMAT(I4, F8.1, 3F9.2)
103     FORMAT(/' Coefficients    Errors')
104     FORMAT(0PF8.3, 3X, 1PE12.3)
105     FORMAT(/' Correlation coefficient is', F7.4)
106     FORMAT(0PF8.3, 3X, 1PE12.3, ' Constant term')

        END
        SUBROUTINE LINFIT(X, Y, YCALC, RESID, COEF, SIG,
     *  NROW, NCOL, COR)
C
C -- Least squares fit to NROW sets of X-Y points
C -- using Gauss-Jordan elimination.
C -- Subroutines SQUARE and GAUSSJ needed.
C
```

Figure 7.5: Input Polynomial Order from the Console (cont.)

```
         LOGICAL ERROR
         INTEGER NROW, NCOL, I, J, MAXR, MAXC, IN, OUT
         INTEGER INDEX(20,3), NVEC
         REAL X(1), Y(1), YCALC(1), COEF(1), RESID(1)
         REAL A(3,3), XMATR(20,3), SIG(1)
         REAL SUMY, SUMY2, XI, YI, YC, RES, COR, SRS, SEE
         COMMON /INOUT/ IN, OUT, MAXR, MAXC
         DATA NVEC/1/
C
         DO 10 I = 1, NROW
           XI = X(I)
           XMATR(I,1) = 1.0
           XMATR(I,2) = XI
           XMATR(I,3) = XI * XI
10       CONTINUE
         CALL SQUARE(XMATR, Y, A, COEF, NROW, NCOL, MAXR, MAXC)
         CALL GAUSSJ(A, COEF, INDEX, NCOL, MAXC, NVEC, ERROR, OUT)
         SUMY = 0.0
         SUMY2 = 0.0
         SRS = 0.0
         DO 20 I = 1, NROW
           YI = Y(I)
           YC = 0.0
           DO 15 J = 1, NCOL
15           YC = YC + COEF(J) * XMATR(I,J)
           YCALC(I) = YC
           RES = YC - YI
           RESID(I) = RES
           SRS = SRS + RES*RES
           SUMY = SUMY + YI
           SUMY2 = SUMY2 + YI * YI
20       CONTINUE
         COR = SQRT(1.0 - SRS/(SUMY2 - SUMY*SUMY/NROW))
         IF (NROW .EQ. NCOL) SEE = SQRT(SRS)
         IF (NROW .NE. NCOL) SEE = SQRT(SRS / (NROW - NCOL))
         DO 30 I = 1, NCOL
30         SIG(I) = SEE * SQRT(A(I,I))
         RETURN
         END
```

Figure 7.5: Input Polynomial Order from the Console (cont.)

Comparing Runs of the Program

Run the program and input a value of 2 for the polynomial order. The results should again be the same as they were for the two previous versions. This time, however, the program will cycle and ask for the polynomial order again. Give a value of 1 the second time. The results (shown in Figure 7.6) are the best straight line through the curved set of data.

```
Order? 1

    I       X          Y        Y CALC      RESID
    1      1.0       2.07       11.79        9.72
    2      2.0       8.60       11.55        2.95
    3      3.0      14.42       11.31       -3.11
    4      4.0      15.80       11.07       -4.73
    5      5.0      18.92       10.83       -8.09
    6      6.0      17.96       10.59       -7.37
    7      7.0      12.98       10.35       -2.63
    8      8.0       6.45       10.11        3.66
    9      9.0        .27        9.87        9.60


Coefficients      Errors
  12.027         5.264E+00   Constant term
   -.239         9.354E-01

Correlation coefficient is   .0963

9 DATA SETS        +-DATA, *-FITTED CURVE, M-MULTIPLE POINT

  1.0000          +                          *
 -
  2.0000                              +       *
 -
  3.0000                                  *        +
 -
  4.0000                                  *            +
 -
  5.0000                                  *                +
 -
  6.0000                                 *             +
 -
  7.0000                                 *        +
 -
  8.0000                     +           *
 -
  9.0000          +                    *
                 ^        ^          ^        ^          ^        ^
               .270    4.000      7.730    11.460     15.190    18.920

                        DEPENDENT VARIABLE
```

Figure 7.6: Output: A Straight-Line Fit to Parabolic Data

The two coefficients represent the equation:

$$y = 12.03 - 0.239x$$

Notice that about half of the data points are on one side of the straight line and half are on the other. The straight line goes through the points, as best it can. The correlation coefficient, however, is less than 0.1. This relatively small value indicates that the straight-line fit is not very good.

With this present version, we can choose a polynomial with an order up to 2. This restriction is necessary because we have set the maximum number of matrix columns to be 3 in the dimension statements. If a higher-order polynomial is necessary, change the value of 3 in each dimension statement to the correspondingly higher value. The value of MAXC must also be changed accordingly.

The program is terminated by entering a polynomial order that is zero or negative. You may want to alter the program so that it asks for the number of terms, rather than the polynomial order. In that case, the statement:

NCOL = NCOL + 1

in the input routine should be removed.

In the next three sections we will look at actual experimental applications of curve fitting involving the heat capacity of oxygen, the vapor pressure of liquid lead, and the properties of superheated steam.

FORTRAN PROGRAM: THE HEAT-CAPACITY EQUATION

Heat capacity is a measure of how much the temperature of a body will increase when a given amount of heat is added. Experimentally determined data are commonly fitted to the equation:

$$C_p = A + BT + \frac{C}{T^2}$$

where C_p is the heat capacity in units of energy per degree, and T is the absolute temperature. The coefficients are A, B, and C, as usual.

Make a copy of the curve-fitting program given in Figure 7.3 (rather than Figure 7.5, which is designed for adjusting the order). Alter subroutines INPUT, OUTPUT, and LINFIT so they look like the versions shown in Figure 7.7. As before, subroutine LINFIT fills the first column of the data matrix with the value of unity, and the second column with the temperature (the independent variable). But now the third column contains the reciprocal of the temperature squared.

Notice that we have changed the variables X and Y in the input subroutine to T and CP. Since these are dummy variables, it is not necessary to change the corresponding names of X and Y in the main program or in the other subroutines.

The data represent the heat capacity of oxygen over the temperature range of 300 to 1200 kelvins. Compile the program and run it. The output should look like Figure 7.8. The resulting equation is:

$$C_p = 6.9 + 0.00143T - \frac{32610}{T^2}$$

```
C       PROGRAM LEAST4
C
C -- Least-squares fit to the heat capacity equation.
C -- Subroutines GAUSSJ, SQUARE, and PLOT required.
C -- May 3, 81
C
        INTEGER IN, OUT, MAXR, MAXC, LINES, NROW, NCOL
        REAL X(20), Y(20), YCALC(20), COEF(3), CORREL
        REAL SIG(3), RESID(20)
        COMMON /INOUT/ IN, OUT, MAXR, MAXC
C
        IN = 1
        OUT = 1
        MAXR = 20
        MAXC = 3
        CALL INPUT(X, Y, NROW)
        CALL LINFIT(X, Y, YCALC, RESID, COEF, SIG,
     *  NROW, NCOL, CORREL)
        CALL OUTPUT(X, Y, YCALC, RESID, COEF, SIG,
     *  NROW, NCOL, CORREL)
        LINES = 2*(NROW - 1) + 1
        CALL PLOT(X, Y, YCALC, NROW, OUT, LINES)
        STOP
        END
        SUBROUTINE INPUT(T, CP, NROW)
C -- Get values for NROW and arrays T and CP.
C
        INTEGER NROW, I
        REAL T(1), CP(1)
C
        NROW = 10
        DO 10 I = 1, NROW
          T(I) = (I + 2) * 100.0
10      CONTINUE
        CP(1) =   7.02
        CP(2) =   7.2
        CP(3) =   7.43
        CP(4) =   7.67
        CP(5) =   7.88
        CP(6) =   8.06
        CP(7) =   8.21
        CP(8) =   8.34
        CP(9) =   8.44
        CP(10) =  8.53
        RETURN
        END
        SUBROUTINE OUTPUT(X, Y, YCALC, RESID, COEF, SIG,
     *  NROW, NCOL, CORREL)
C
C -- Print out the answers.
C
        INTEGER IN, OUT, NROW, NCOL, I
        REAL X(1), Y(1), YCALC(1), COEF(1), CORREL
```

Figure 7.7: Least-Squares Fit to the Heat-Capacity Equation

```
        REAL RESID(1), SIG(1)
        COMMON /INOUT/ IN, OUT
C
        WRITE(OUT, 101)
        WRITE(OUT, 102) (I, X(I), Y(I), YCALC(I),
     *   RESID(I), I = 1, NROW)
        WRITE(OUT, 103)
        WRITE(OUT, 106) COEF(1), SIG(1)
        WRITE(OUT, 104) (COEF(I), SIG(I), I = 2, NCOL)
        WRITE(OUT, 105) CORREL
        RETURN
101     FORMAT('   I        X         Y       Y CALC     RESID')
102     FORMAT(I4, F8.1, 3F9.2)
103     FORMAT(/' Coefficients     Errors')
104     FORMAT(1X, 1PE12.3, 3X, E12.3)
105     FORMAT(/' Correlation coefficient is', F7.4)
106     FORMAT(1X, 1PE12.3, 3X, E12.3, '  Constant term')
        END
        SUBROUTINE LINFIT(X, Y, YCALC, RESID, COEF, SIG,
     *   NROW, NCOL, COR)
C
C -- Least squares fit to NROW sets of X-Y points
C -- for heat capacity equation
C -- using Gauss-Jordan elimination.
C -- Subroutines SQUARE and GAUSSJ needed.
C
        LOGICAL ERROR
        INTEGER NROW, NCOL, I, J, MAXR, MAXC, IN, OUT
        INTEGER INDEX(20,3), NVEC
        REAL X(1), Y(1), YCALC(1), COEF(1), RESID(1)
        REAL A(3,3), XMATR(20,3), SIG(1)
        REAL SUMY, SUMY2, XI, YI, YC, RES, COR, SRS, SEE
        COMMON /INOUT/ IN, OUT, MAXR, MAXC
        DATA NVEC/1/
C
        NCOL = 3
        DO 10 I = 1, NROW
          XI = X(I)
          XMATR(I,1) = 1.0
          XMATR(I,2) = XI
          XMATR(I,3) = 1.0 / (XI * XI)
10      CONTINUE
        CALL SQUARE(XMATR, Y, A, COEF, NROW, NCOL, MAXR, MAXC)
        CALL GAUSSJ(A, COEF, INDEX, NCOL, MAXC, NVEC, ERROR, OUT)
        SUMY = 0.0
        SUMY2 = 0.0
        SRS = 0.0
        DO 20 I = 1, NROW
          YI = Y(I)
          YC = 0.0
          DO 15 J = 1, NCOL
15          YC = YC + COEF(J) * XMATR(I,J)
```

Figure 7.7: Least-Squares Fit to the Heat-Capacity Equation (cont.)

```
           YCALC(I) = YC
           RES = YC - YI
           RESID(I) = RES
         SRS = SRS + RES*RES
         SUMY = SUMY + YI
         SUMY2 = SUMY2 + YI * YI
20       CONTINUE
         COR = SQRT(1.0 - SRS/(SUMY2 - SUMY*SUMY/NROW))
         IF (NROW .EQ. NCOL) SEE = SQRT(SRS)
         IF (NROW .NE. NCOL) SEE = SQRT(SRS / (NROW - NCOL))
         DO 30 I = 1, NCOL
30       SIG(I) = SEE * SQRT(A(I,I))
         RETURN
         END
```

Figure 7.7: Least-Squares Fit to the Heat-Capacity Equation (cont.)

```
   I      X        Y      Y CALC    RESID
   1    300.0     7.02     6.97     -.05
   2    400.0     7.20     7.27      .07
   3    500.0     7.43     7.49      .06
   4    600.0     7.67     7.67      .00
   5    700.0     7.88     7.84     -.04
   6    800.0     8.06     8.00     -.06
   7    900.0     8.21     8.16     -.05
   8   1000.0     8.34     8.31     -.03
   9   1100.0     8.44     8.46      .02
  10   1200.0     8.53     8.60      .07

Coefficients      Errors
   6.905E+00       1.292E-01   Constant term
   1.434E-03       1.268E-04
  -3.261E+04       1.167E+04

Correlation coefficient is  .9948


10 DATA SETS      +-DATA, *-FITTED CURVE, M-MULTIPLE POINT

 300.0000     **
-
 400.0000         +    *
-
 500.0000               +*
-
 600.0000                  *
-
 700.0000                       +*
-
 800.0000                          *  +
-
 900.0000                             +*
-
1000.0000                                +*
-
1100.0000                                   +*
-
1200.0000                                      +   *
            6.973   7.299   7.625   7.951   8.277   8.604

                    DEPENDENT VARIABLE
```

Figure 7.8: Output: The Heat Capacity of Oxygen

FORTRAN PROGRAM: THE VAPOR PRESSURE EQUATION

When a gas or vapor is in equilibrium with its own liquid or solid, we say that it is saturated. For pure materials, the saturation pressure is a single-valued function of the temperature. A commonly used equation to express the relationship between the saturation pressure and the saturation temperature is:

$$\log P = A + \frac{B}{T} + C \log T$$

In this equation, P is the pressure and T is the absolute temperature. The coefficients are A, B, and C. Either the natural logarithm or the common logarithm is used.

Make a copy of the program shown in Figure 7.7. Alter subroutines INPUT, OUTPUT, and LINFIT so they look like the versions shown in Figure 7.9. Subroutine LINFIT sets the first column of the matrix to unity as usual. Column 2 is then filled with the reciprocal of the independent variable, the temperature. Column 3 gets the logarithm of the temperature. Run the new program. The results should look like Figure 7.10.

```
C        PROGRAM LEAST5
C
C -- Least-squares fit to the vapor pressure of lead.
C -- Subroutines GAUSSJ and SQUARE required.
C -- May 3, 81
C
         INTEGER IN, OUT, MAXR, MAXC, LINES, NROW, NCOL
         REAL X(20), Y(20), YCALC(20), COEF(3), CORREL
         REAL SIG(3), RESID(20)
         COMMON /INOUT/ IN, OUT, MAXR, MAXC
C
         IN = 1
         OUT = 1
         MAXR = 20
         MAXC = 3
         CALL INPUT(X, Y, NROW)
         CALL LINFIT(X, Y, YCALC, RESID, COEF, SIG,
      *  NROW, NCOL, CORREL)
         CALL OUTPUT(X, Y, YCALC, RESID, COEF, SIG,
      *  NROW, NCOL, CORREL)
         STOP
         END
         SUBROUTINE INPUT(T, P, NROW)
C -- Get values for NROW and arrays T and P.
C
         INTEGER NROW, I
         REAL T(1), P(1)
```

Figure 7.9: Least-Squares Fit to the Vapor Pressure Equation

```
C
        NROW = 10
        DO 10 I = 1, NROW
           T(I) = (I + 6) * 100.0
10      CONTINUE
        P(1) =   1.00E-9
        P(2) =   5.598E-8
        P(3) =   1.234E-6
        P(4) =   1.507E-5
        P(5) =   1.138E-4
        P(6) =   6.067E-4
        P(7) =   2.512E-3
        P(8) =   8.337E-3
        P(9) =   2.371E-2
        P(10) =  5.875E-2
        DO 20 I = 1,NROW
           P(I) = ALOG(P(I))
20      CONTINUE
        RETURN
        END
        SUBROUTINE OUTPUT(X, Y, YCALC, RESID, COEF, SIG,
     *  NROW, NCOL, CORREL)
C
C -- Print out the answers.
C
        INTEGER IN, OUT, NROW, NCOL, I
        REAL X(1), Y(1), YCALC(1), COEF(1), CORREL
        REAL RESID(1), SIG(1)
        COMMON /INOUT/ IN, OUT
C
        WRITE(OUT, 101)
        WRITE(OUT, 102) (I, X(I), Y(I), YCALC(I),
     *  RESID(I), I = 1, NROW)
        WRITE(OUT, 103)
        WRITE(OUT, 106) COEF(1), SIG(1)
        WRITE(OUT, 104) (COEF(I), SIG(I), I = 2, NCOL)
        WRITE(OUT, 105) CORREL
        RETURN
101     FORMAT('   I      X         Y      Y CALC     RESID')
102     FORMAT(I4, F8.0, 3F9.2)
103     FORMAT(/' Coefficients      Errors')
104     FORMAT(1X, 1PE12.3, 3X, E12.3)
105     FORMAT(/' Correlation coefficient is', F7.4)
106     FORMAT(1X, 1PE12.3, 3X, E12.3, '  Constant term')
        END
        SUBROUTINE LINFIT(X, Y, YCALC, RESID, COEF, SIG,
     *  NROW, NCOL, COR)
C
C -- Least squares fit to NROW sets of X-Y points
C -- for vapor pressure equation
C -- using Gauss-Jordan elimination.
C -- Subroutines SQUARE and GAUSSJ needed.
C
```

Figure 7.9: Least-Squares Fit to the Vapor Pressure Equation (cont.)

```
      LOGICAL ERROR
      INTEGER NROW, NCOL, I, J, MAXR, MAXC, IN, OUT
      INTEGER INDEX(20,3), NVEC
      REAL X(1), Y(1), YCALC(1), COEF(1), RESID(1)
      REAL A(3,3), XMATR(20,3), SIG(1)
      REAL SUMY, SUMY2, XI, YI, YC, RES, COR, SRS, SEE
      COMMON /INOUT/ IN, OUT, MAXR, MAXC
      DATA NVEC/1/
C
      NCOL = 3
      DO 10 I = 1, NROW
        XI = X(I)
        XMATR(I,1) = 1.0
        XMATR(I,2) = 1.0 / XI
        XMATR(I,3) = ALOG(XI)
10    CONTINUE
      CALL SQUARE(XMATR, Y, A, COEF, NROW, NCOL, MAXR, MAXC)
      CALL GAUSSJ(A, COEF, INDEX, NCOL, MAXC, NVEC, ERROR, OUT)
      SUMY = 0.0
      SUMY2 = 0.0
      SRS = 0.0
      DO 20 I = 1, NROW
        YI = Y(I)
        YC = 0.0
        DO 15 J = 1, NCOL
15         YC = YC + COEF(J) * XMATR(I,J)
        YCALC(I) = YC
        RES = YC - YI
        RESID(I) = RES
        SRS = SRS + RES*RES
        SUMY = SUMY + YI
        SUMY2 = SUMY2 + YI * YI
20    CONTINUE
      COR = SQRT(1.0 - SRS/(SUMY2 - SUMY*SUMY/NROW))
      IF (NROW .EQ. NCOL) SEE = SQRT(SRS)
      IF (NROW .NE. NCOL) SEE = SQRT(SRS / (NROW - NCOL))
      DO 30 I = 1, NCOL
30       SIG(I) = SEE * SQRT(A(I,I))
      RETURN
      END
```

Figure 7.9: Least-Squares Fit to the Vapor Pressure Equation (cont.)

The data represent the vapor pressure of liquid lead over the temperature range of 700 to 1600 kelvins. The corresponding equation is:

$$\ln P = 18 - \frac{23160}{T} - .869 \ln T$$

when the pressure is in atmospheres and the natural logarithm is used. The correlation coefficient of unity indicates an excellent fit.

```
    I        X          Y        Y CALC      RESID
    1      700.     -20.72      -20.72       -.00
    2      800.     -16.70      -16.70       -.01
    3      900.     -13.61      -13.59        .02
    4     1000.     -11.10      -11.11       -.01
    5     1100.      -9.08       -9.09       -.00
    6     1200.      -7.41       -7.41        .00
    7     1300.      -5.99       -5.99       -.01
    8     1400.      -4.79       -4.78        .00
    9     1500.      -3.74       -3.74        .00
   10     1600.      -2.83       -2.83        .00

  Coefficients          Errors
     1.805E+01          5.948E-01   Constant term
    -2.316E+04          7.751E+01
    -8.690E-01          7.457E-02

  Correlation coefficient is 1.0000
```

Figure 7.10: Output: The Vapor Pressure of Lead

Notice that the original pressures in the input subroutine are given in atmospheres. These values are then converted to the logarithm of the pressure. The X and Y values that are printed are actually the temperature and the logarithm of the pressure. It is left as an exercise for the reader to include the original pressure values in the final printout.

In the final example we will encounter an equation that breaks the restriction we established at the beginning of the chapter. One of the unknown coefficients is in a nonlinear form—specifically, an exponential. Later, in Chapter 10, we will study algorithms for handling such an equation, but here we will have to be satisfied with a less elegant solution. We will estimate values for the exponent until we find the optimum correlation coefficient.

A THREE-VARIABLE EQUATION

In the previous section, we considered an equation of state for a saturated gas. The pressure is a function of temperature for this condition. We will now consider an equation of state for a superheated gas. In this case, the temperature is above the saturation temperature or, put another way, the pressure is below the saturation pressure. Since temperature and pressure are independent variables under this condition, we can express the volume as a function of both the temperature and the pressure.

For an ideal gas, the equation of state is:

$$PV = RT$$

where P is the pressure, V is the molar volume, R is the gas constant, and T is the temperature. But as the pressure increases, the behavior of the gas becomes less and less ideal. A common non-ideal equation of state is:

$$PV = A + BP + CP^2 + DP^3 + ...$$

It can be seen that this equation is just a power series expansion in pressure.

Be careful not to confuse the gas constant (R) with the vector of residuals ($\mathbf{r}$). In the FORTRAN program, the symbol GAS is used for the gas constant and the symbol RESID is used for the vector of residuals.

Since all gases become ideal as the pressure is reduced, the equation of state should merge smoothly with the ideal gas equation as the pressure approaches zero. Thus, the value of A in the above equation must be equal to RT. Because fewer polynomial terms are needed at lower values of pressure, the equation of state can often be written as:

$$PV = RT + BP + CP^2$$

The determination of the coefficients B and C is straightforward when the temperature is constant. However, if temperature is also a variable, then the coefficients B and C may need to be functions of temperature.

Several different equations of state, all of them empirical, are in common use. As an example of a three-variable equation, consider the expression:

$$PV = RT + \frac{BP}{T^n} + CP^2$$

In this example, coefficient C is not a function of T, but the original coefficient B has become the function:

$$\frac{B}{T^n}$$

This equation has a nonlinear coefficient, n, and so we cannot obtain a solution by the methods discussed in this chapter. However, if an estimate is made for the coefficient n, then the remaining linear coefficients can be determined. We will begin with an estimate of unity for coefficient n and determine the other coefficients by our usual least-squares method. We can then observe how well the resulting equation represents the original data. Then we will change the value of n and see whether the new equation becomes better or worse.

FORTRAN PROGRAM: AN EQUATION OF STATE FOR STEAM

The program given in Figure 7.11 can be used to find the coefficients B and C for the published properties of steam. Since the coefficient of the first term on the right is unity, the equation has been rearranged to give:

$$PV - RT = \frac{BP}{T^n} + CP^2$$

The data are defined in the input routine as usual. Temperature is given in degrees Fahrenheit, the pressure in pounds per square inch, and the specific volume in cubic feet per pound mass. The temperature data are converted to the absolute Rankine scale by the addition of 460. Pressures are left in pounds per square inch. But then the gas constant, *R,* which has a value of 85.76 for steam, is divided by 144 square inches per square foot.

The data matrix is set up in subroutine LINFIT as usual. The first column of the matrix contains the pressure divided by the *n*th power of the temperature. The second column contains the square of the pressure. The **y** vector has the value *PV − RT.*

Notice that this is the first time that we did *not* put the value of unity in the first column of the matrix. We could, of course, divide the equation by the pressure. The right-hand side would then look like a first-order, straight-line fit. However, dividing the equation by the pressure might cause trouble when the pressure became very small.

Comparing Runs of the Program to Determine the Value of *n*

Type up the program given in Figure 7.11 and execute it. The program will ask for the value of the exponent *n.* Give a value of unity for the first run. The results should look like Figure 7.12.

```
C        PROGRAM LEAST6
C
C -- Least-squares fit to p-v-t of steam.
C -- Subroutines GAUSSJ and SQUARE required.
C -- May 3, 81
C
         INTEGER IN, OUT, MAXR, MAXC, NROW, NCOL
         REAL Y(20), YCALC(20), COEF(3), CORREL
         REAL SIG(3), RESID(20), P(29), T(20), V(20)
         COMMON /INOUT/ IN, OUT, MAXR, MAXC
C
         IN = 1
         OUT = 1
         MAXR = 20
         MAXC = 3
         CALL INPUT(P, T, V, NROW)
10       CALL LINFIT(P, T, V, Y, YCALC, RESID, COEF, SIG,
     *   NROW, NCOL, CORREL)
         CALL OUTPUT(P, T, V, Y, YCALC, RESID, COEF, SIG,
     *   NROW, NCOL, CORREL)
         GOTO 10
         END
         SUBROUTINE INPUT(P, T, V, NROW)
C -- Get values for NROW and arrays T and P.
C
         INTEGER NROW, I
```

Figure 7.11: An Equation of State for Steam

```
          REAL T(1), P(1), V(1)
C
          NROW = 12
          T(1) = 400
          P(1) = 120
          V(1) = 4.079
          T(2) = 450
          P(2) = 120
          V(2) = 4.36
          T(3) = 500
          P(3) = 120
          V(3) = 4.633
          T(4) = 400
          P(4) = 140
          V(4) = 3.466
          T(5) = 450
          P(5) = 140
          V(5) = 3.713
          T(6) = 500
          P(6) = 140
          V(6) = 3.952
          T(7) = 400
          P(7) = 160
          V(7) = 3.007
          T(8) = 450
          P(8) = 160
          V(8) = 3.228
          T(9) = 500
          P(9) = 160
          V(9) = 3.44
          T(10) = 400
          P(10) = 180
          V(10) = 2.648
          T(11) = 450
          P(11) = 180
          V(11) = 2.85
          T(12) = 500
          P(12) = 180
          V(12) = 3.042
C -- Convert T to Rankine.
          DO 20 I = 1, NROW
             T(I) = T(I) + 460
20        CONTINUE
          RETURN
          END
          SUBROUTINE OUTPUT(P, T, V, Y, YCALC, RESID, COEF, SIG,
     *    NROW, NCOL, CORREL)
C
C -- Print out the answers.
C
          INTEGER IN, OUT, NROW, NCOL, I
          REAL P(1), T(1), V(1), Y(1), YCALC(1), COEF(1), CORREL
```

Figure 7.11: An Equation of State for Steam (cont.)

```
          REAL RESID(1), SIG(1), RES(20)
          COMMON /INOUT/ IN, OUT
C
          WRITE(OUT, 101)
          DO 10 I = 1, NROW
             RES(I) = 100 * RESID(I) / Y(I)
10        CONTINUE
          WRITE(OUT, 102) (I, P(I), T(I), V(I), Y(I),
       *    YCALC(I), RES(I), I = 1, NROW)
          WRITE(OUT, 103)
          WRITE(OUT, 106) COEF(1), SIG(1)
          WRITE(OUT, 104) (COEF(I), SIG(I), I = 2, NCOL)
          WRITE(OUT, 105) CORREL
          RETURN
101       FORMAT('  I     P       T       V        Y',
       *    '      Y CALC    % RES')
102       FORMAT(I4, 2F7.0, F7.3, 3F9.2)
103       FORMAT(/' Coefficients       Errors')
104       FORMAT(1X, 1PE12.3, 3X, E12.3)
105       FORMAT(/' Correlation coefficient is', F7.4)
106       FORMAT(1X, 1PE12.3, 3X, E12.3, '  Constant term')
          END
          SUBROUTINE LINFIT(P, T, V, Y, YCALC, RESID, COEF, SIG,
       *    NROW, NCOL, COR)
C
C -- Least squares fit to NROW sets of X-Y points
C -- for properties of steam
C -- using Gauss-Jordan elimination.
C -- Subroutines SQUARE and GAUSSJ needed.
C
          LOGICAL ERROR
          INTEGER NROW, NCOL, I, J, MAXR, MAXC, IN, OUT
          INTEGER INDEX(20,3), NVEC
          REAL P(1), T(1), V(1), Y(1), YCALC(1), COEF(1), RESID(1)
          REAL A(3,3), XMATR(20,3), SIG(1), GAS, POWER
          REAL SUMY, SUMY2, XI, YI, YC, RES, COR, SRS, SEE
          COMMON /INOUT/ IN, OUT, MAXR, MAXC
          DATA NVEC/1/, GAS/85.76/
C
          NCOL = 2
          WRITE(OUT, 101)
          READ(IN, 102) POWER
          IF (POWER .LT. -10.0) GOTO 99
          DO 10 I = 1, NROW
             XMATR(I,1) = P(I) / T(I)**POWER
             XMATR(I,2) = SQRT(P(I))
             Y(I) = V(I) * P(I) - GAS * T(I) / 144
10        CONTINUE
          CALL SQUARE(XMATR, Y, A, COEF, NROW, NCOL, MAXR, MAXC)
          CALL GAUSSJ(A, COEF, INDEX, NCOL, MAXC, NVEC, ERROR, OUT)
          SUMY = 0.0
          SUMY2 = 0.0
          SRS = 0.0
```

Figure 7.11: An Equation of State for Steam (cont.)

```
         DO 20 I = 1, NROW
           YI = Y(I)
           YC = 0.0
           DO 15 J = 1, NCOL
15           YC = YC + COEF(J) * XMATR(I,J)
           YCALC(I) = YC
           RES = YC - YI
           RESID(I) = RES
           SRS = SRS + RES*RES
           SUMY = SUMY + YI
           SUMY2 = SUMY2 + YI * YI
20       CONTINUE
         COR = SQRT(1.0 - SRS/(SUMY2 - SUMY*SUMY/NROW))
         IF (NROW .EQ. NCOL) SEE = SQRT(SRS)
         IF (NROW .NE. NCOL) SEE = SQRT(SRS / (NROW - NCOL))
         DO 30 I = 1, NCOL
30         SIG(I) = SEE * SQRT(A(I,I))
         RETURN
99       STOP
101      FORMAT(/' Power? ')
102      FORMAT(E10.0)
         END
```

Figure 7.11: An Equation of State for Steam (cont.)

```
Power? 1
    I     P      T      V        Y      Y CALC     % RES
    1    120.   860.  4.079   -22.70   -19.44    -14.36
    2    120.   910.  4.360   -18.76   -17.43     -7.07
    3    120.   960.  4.633   -15.77   -15.63      -.92
    4    140.   860.  3.466   -26.94   -24.16    -10.30
    5    140.   910.  3.713   -22.14   -21.82     -1.44
    6    140.   960.  3.952   -18.45   -19.72      6.85
    7    160.   860.  3.007   -31.06   -28.98     -6.69
    8    160.   910.  3.228   -25.48   -26.30      3.24
    9    160.   960.  3.440   -21.33   -23.90     12.03
   10    180.   860.  2.648   -35.54   -33.88     -4.67
   11    180.   910.  2.850   -28.96   -30.86      6.59
   12    180.   960.  3.042   -24.17   -28.16     16.50

  Coefficients       Errors
   -2.622E+02       4.759E+01   Constant term
    1.565E+00       6.495E-01

Correlation coefficient is  .9184
```

Figure 7.12:
Output: The Properties of Superheated Steam (The exponent n = 1.)

It can be seen from Figure 7.12 that the results are not very good. The correlation coefficient is 92%, but some of the calculated values are more than 10% from the original data. We should therefore try something else. Enter a value of 2 for the exponent. The output from the second run should look like Figure 7.13.

```
Power? 2

   I      P        T        V           Y         Y CALC      % RES
   1    120.     860.    4.079      -22.70      -21.49       -5.32
   2    120.     910.    4.360      -18.76      -18.03       -3.85
   3    120.     960.    4.633      -15.77      -15.10       -4.25
   4    140.     860.    3.466      -26.94      -26.01       -3.44
   5    140.     910.    3.713      -22.14      -21.98        -.71
   6    140.     960.    3.952      -18.45      -18.56         .58
   7    160.     860.    3.007      -31.06      -30.59       -1.49
   8    160.     910.    3.228      -25.48      -25.98        2.00
   9    160.     960.    3.440      -21.33      -22.08        3.48
  10    180.     860.    2.648      -35.54      -35.22        -.88
  11    180.     910.    2.850      -28.96      -30.04        3.74
  12    180.     960.    3.042      -24.17      -25.64        6.08

  Coefficients          Errors
   -1.994E+05          1.194E+04   Constant term
    9.909E-01          1.803E-01

  Correlation coefficient is   .9890
```

Figure 7.13:
Output: The Properties of Superheated Steam (The exponent n = 2.)

The resulting curve fit looks better this time. The correlation coefficient is 99%, and the calculated points are within about 6% of the data. We will now continue in this way by increasing the exponent to 3. Run the program again and compare the output this time to Figure 7.14.

```
Power? 3

   I      P        T        V           Y         Y CALC      % RES
   1    120.     860.    4.079      -22.70      -22.88         .78
   2    120.     910.    4.360      -18.76      -18.80         .22
   3    120.     960.    4.633      -15.77      -15.52       -1.59
   4    140.     860.    3.466      -26.94      -26.97         .13
   5    140.     910.    3.713      -22.14      -22.21         .35
   6    140.     960.    3.952      -18.45      -18.39        -.32
   7    160.     860.    3.007      -31.06      -31.09         .10
   8    160.     910.    3.228      -25.48      -25.65         .68
   9    160.     960.    3.440      -21.33      -21.28        -.23
  10    180.     860.    2.648      -35.54      -35.22        -.90
  11    180.     910.    2.850      -28.96      -29.10         .49
  12    180.     960.    3.042      -24.17      -24.19         .06

  Coefficients          Errors
   -1.387E+08          1.502E+06   Constant term
    2.998E-01          2.520E-02

  Correlation coefficient is   .9996
```

Figure 7.14:
Output: The Properties of Superheated Steam (The exponent n = 3.)

The results are definitely better. The fitted values are within 2% of the original points and the correlation coefficient is 99.96%. It looks as though we should accept these results. But just to be sure, try an exponent of 4. Compare the results to Figure 7.15. It can be seen that we have definitely gone too far. The calculated points are not as close as they were for the previous fit and the correlation coefficient is farther from unity. In fact, the Gauss-Jordan procedure may report that the matrix is singular. The problem is ill conditioning. This can be seen by comparing the squares of the two standard errors. This ratio is greater than twenty orders of magnitude.

At this point, we could try to improve the fit by choosing noninteger exponents close to the value of 3. The results would show that 3 is best. Another possibility is to make coefficient C a function of temperature. But since the result with an exponent of 3 is reasonably good, we should perhaps be satisfied. Give a power that is less than -10 to abort the program.

```
Power?  4

  I      P        T        V         Y       Y CALC    % RES
  1    120.     860.     4.079    -22.70    -23.69     4.38
  2    120.     910.     4.360    -18.76    -19.32     3.02
  3    120.     960.     4.633    -15.77    -16.00     1.45
  4    140.     860.     3.466    -26.94    -27.46     1.94
  5    140.     910.     3.713    -22.14    -22.36     1.02
  6    140.     960.     3.952    -18.45    -18.49      .19
  7    160.     860.     3.007    -31.06    -31.22      .51
  8    160.     910.     3.228    -25.48    -25.39     -.34
  9    160.     960.     3.440    -21.33    -20.96    -1.74
 10    180.     860.     2.648    -35.54    -34.96    -1.62
 11    180.     910.     2.850    -28.96    -28.41    -1.89
 12    180.     960.     3.042    -24.17    -23.43    -3.08

  Coefficients         Errors
  -9.847E+10           3.648E+09   Constant term
  -1.908E-01           6.829E-02

  Correlation coefficient is  .9957
```

Figure 7.15:
Output: The Properties of Superheated Steam (The exponent n = 4.)

SUMMARY

This chapter has illustrated the programming concept of *generalization*. We have now progressed through several versions of our curve-fitting program:

- a straight-line fit

- a parabolic fit

- a more direct parabolic fit, using the matrix approach

- a general polynomial curve fit in which the polynomial order can be adjusted

- curve fits for nonpolynomial equations.

The restriction applying to all of these versions is that the coefficients must be linear. Although we found a way around this restriction in the last example of the chapter, we must still develop a general method of dealing with nonlinear coefficients. We will return to this topic in Chapter 10.

EXERCISES

7-1: *It is possible to calculate the thermodynamic activity of one component of a binary solution from a knowledge of the activity of the other component. The common logarithm of the activity coefficient, G, (the ratio of activity to mole fraction) is fitted to a power series in mole fraction of the other component. Notice that the series begins with the second-order term.*

$$\text{Log } G_1 = A\, x_2^2 + B\, x_2^3 + C\, x_2^4 + \ldots$$

The activity, A_{Ni}, of nickel in iron-nickel solutions at 1500 kelvins as a function of the mole fraction of nickel, X_{Ni}, has been reported as:

X_{Ni}	1	.9	.8	.7	.6	.5	.4	.3	.2	.1
A	1	.89	.766	.62	.485	.374	.283	.207	.136	.067

Make a copy of the program given in Figure 7.3. Increase the number of data points, NROW, to 10 in subroutine INPUT. Change the definition of X so that it represents the mole fraction of iron:

X(I) = 0.1 * (I − 1)

Change the following values of Y so that they correspond to the nickel activities. Change subroutine LINFIT so that the power series begins with the second-order term:

XMATR(I, 1) = X(I) * X(I)

Change the dependent variable with the statement:

Y(I) = ALOG(Y(I) / (1.0 − X(I)))

In this step, the activity coefficient is obtained from a ratio of the activity to the mole fraction. The dependent variable is the logarithm of the ratio. Determine the coefficients A, B, and C.

Answer: A = −2.128, B = 2.319, C = −0.548

7-2: The vapor pressure of mercury is reported to be:

Temperature, C	Pressure, torr
0	0.000185
50	0.01267
100	0.2729
150	2.2807
200	17.287
250	74.375
300	246.80
350	672.69
400	1574.1

Perform a least-squares fit on the data to find the coefficients A, B, and C to the equation:

$$\ln P = A + B/T + C \ln T$$

The x data can be generated in the input loop:

X(I) = (I − 1) * 50 + 273.15

Define the y data as the above pressure values, then convert them with the statement:

Y(I) = ALOG(Y(I))

Use the vapor pressure program developed in Figure 7.9.

Answer: A = 20.57, B = 7469, C = 0.3226

7-3: Find the coefficients A, B, and C to the expression:

$$C_p = A + BT + C/T^2$$

for the heat capacity of graphite. Use the program given in Figure 7.7. The temperatures are given in kelvins and the heat capacities are given in cal/deg mole.

T	C_p
300	2.08
400	2.85
500	3.50
600	4.03
700	4.43
800	4.75
900	4.98
1000	5.14
1100	5.27
1200	5.42

Answer: A = 3.37, B = 0.00198, C = −177100

8

Solution of Equations by Newton's Method

IN THIS CHAPTER, we will develop a computer program for solving equations by a technique known as Newton's method, or the Newton-Raphson method. This technique is especially suitable for finding particular roots of "well-behaved" functions. We will use Newton's method in Chapter 10 when we return to the topic of curve fitting with nonlinear coefficients. Of course, as we will see, Newton's method has many other applications and would be worth our attention even if we did not plan to use it as a tool later in this book.

First we will outline the mathematical formulation of Newton's method; then we will consider a series of progressively more significant FORTRAN implementations of the technique. We will study two pitfalls of the method—the case where the tangent to the curve has a zero slope, and the case where successive approximations fail to converge on a root. We will then consider ways of dealing with these pitfalls in our program. Finally, we will use our program to solve a practical application—the vapor pressure equation.

FORMULATING NEWTON'S METHOD

Let us begin by considering a general equation of the form:

$$f(x) = 0 \tag{1}$$

This equation might have one solution, several solutions, or none at all. That is, there may be particular values of x that make the equation equal to zero. These values are called the *roots* or *solutions* of the equation. For other values of x, the function will not be zero.

Sometimes we can solve such an equation explicitly. For example, the expression:

$$x^2 - 4 = 0$$

can be converted to:

$$x^2 = 4$$

which has the solutions:

$$x = 2 \quad \text{and} \quad x = -2$$

But sometimes, an equation cannot be solved so easily. As an example, consider the expression:

$$\ln P = A + \frac{B}{T} + C \ln T$$

This formula, which we will implement into our program at the end of this chapter, can be used to describe the vapor pressure of a material. In this equation P is the pressure, T is the temperature, and A, B, and C are constants that are unique for each substance. For the element lead, the experimentally determined coefficients are:

$$A = 18.19$$
$$B = -23180$$
$$C = -0.8858$$

when the pressure is given in atmospheres and the temperature is given in kelvins.

We can easily find the vapor pressure of lead at, say, 1000°K by solving the expression:

$$\ln P = A + \frac{B}{1000} + C \ln 1000$$

But suppose that we want to find the temperature that corresponds to a lead vapor pressure of 0.1 atmospheres. We want to solve the equation:

$$\ln 0.1 = 18.19 - \frac{23180}{T} - 0.8858 \ln T$$

This nonlinear equation cannot be solved explicitly for the temperature. However, we can use an approximation method to calculate the answer to as high a precision as we desire.

For the general case, we write the equation:

$$y = f(x)$$

We are thus interested in the values of x when y is equal to zero; that is, we want to determine the points where the curve of the function crosses the x-axis.

Consider, for example, the curve of the equation:

$$y = f(x) = x^2 - 4$$

This curve crosses the x-axis at two places, $+2$ and -2, as shown in Figure 8.1.

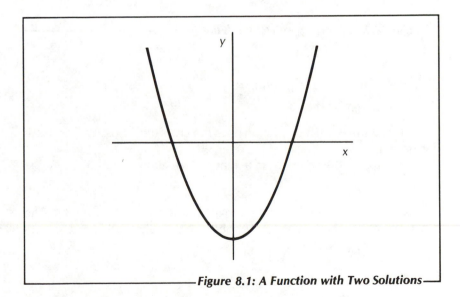

Figure 8.1: A Function with Two Solutions

As a second example, consider the curve of:

$$y = f(x^2) = x^2$$

This curve is tangent to the x-axis at the origin, corresponding to a single root, $x = 0$, as shown in Figure 8.2.

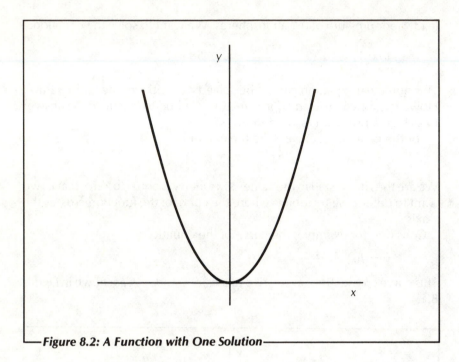

Figure 8.2: A Function with One Solution

Finally, consider the equation:

$$y = f(x) = x^2 + 4$$

shown in Figure 8.3. This equation does not cross the x-axis at all, and so it has no real roots.

Let us explore the behavior of a general function,

$$y = f(x)$$

near a root. We might find that it looks like the curve of Figure 8.4. The function crosses the x-axis at a root because the relationship:

$$y = f(x) = 0$$

is satisfied there.

A Series of Tangents to the Curve y = f(x)

We start Newton's method with an approximate value for x, say x_1, that is near a root. We can determine the corresponding value of y by the equation $y_1 = f(x_1)$. This will represent a point on the curve that is not, in general, a root. A tangent to $f(x)$ is now constructed at this point on the

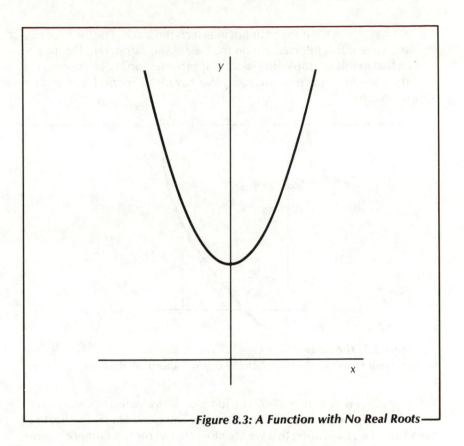

Figure 8.3: A Function with No Real Roots

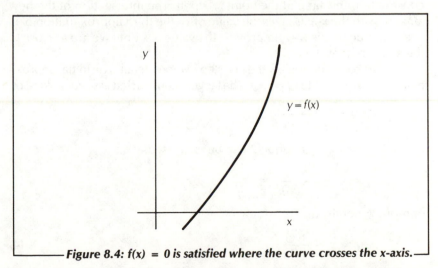

Figure 8.4: f(x) = 0 is satisfied where the curve crosses the x-axis.

curve. The tangent is extended until it intersects the x-axis. The next approximation, x_2, is at this intersection on the x-axis, as illustrated in Figure 8.5. Notice that in this example, the second approximation, x_2, is closer to the root than the first approximation, x_1. We have thus refined our original approximation.

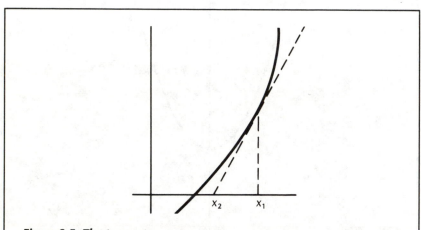

Figure 8.5: The tangent crosses the x-axis closer to the root than the original approximation for x.

The process is now repeated. The function is evaluated at $x = x_2$ to obtain $y_2 = f(x_2)$, the corresponding value of y. The value of y_2 is smaller than the value of y_1, indicating that we are closer to the root. A tangent is again constructed, this time at the point $(x_2, f(x_2))$. The intersection of the new tangent with the x-axis gives the value of x_3, the third approximation of x. We continue in this way, improving the value of x until we are as close to the actual root as we want.

Let us go back and review the first step in more detail. The initial approximation, x_1, gives rise to $y_1 = f(x_1)$. The tangent constructed at y_1 has a slope of:

$$f'(x_1) = \frac{y_1}{x_1 - x_2} \qquad (2)$$

Because $y_1 = f(x_1)$, Equation 2 can be expressed as:

$$x_2 = x_1 - \frac{f(x_1)}{f'(x_1)} \qquad (3)$$

or more generally as:

$$x_{i+1} = x_i - \frac{f(x_i)}{f'(x_i)} \qquad (4)$$

where x_i is the ith approximation. Equation 4 is the usual form of Newton's method. It can be an ideal technique for finding a desired root for a real-life equation.

There are potential problems with the use of Newton's method, which we will consider later in this chapter. But the equations that deal with the behavior of real things typically have only one meaningful root. The other roots of such equations will usually be negative, zero, or complex. Furthermore, the approximate value of the answer may be known. For example, the ideal-gas law can provide a first approximation to a more complicated equation of state.

Now that we have arrived at an equation for the general form of Newton's method, developing our program will be relatively easy.

FORTRAN PROGRAM: A FIRST ATTEMPT AT NEWTON'S METHOD

We will implement Newton's method for a simple problem, one for which we already know the answer. The equation we will solve is:

$$x^2 = 2$$

or

$$x^2 - 2 = 0 \tag{5}$$

The positive solution to this equation is, of course, the square root of 2. First, we define the function:

$$y = f(x) = x^2 - 2 \tag{6}$$

and its derivative:

$$\frac{dy}{dx} = f'(x) = 2x \tag{7}$$

Our first attempt at a Newton's method program is shown in Figure 8.6. The algorithm itself is contained in subroutine NEWTON. A separate subroutine, called FUNC, is used for the calculation of the function $f(x)$ (Equation 6) and its derivative $f'(x)$ (Equation 7). The first approximation is established at the beginning of the main program. Then the Newton's method subroutine is called to find the solution and to print out the answer.

As a matter of good programming practice, subroutines and functions that are used to calculate values should not print intermediate results. But in this particular case, it is instructive to observe the successive values of x, $f(x)$ and $f'(x)$ during the convergence process. Consequently, there are two WRITE statements in subroutine NEWTON for this purpose. You may want to remove these statements and their corresponding format statements when the program is working properly.

```
C       PROGRAM NEWT1
C
        INTEGER IN, OUT
        REAL X
        COMMON /INOUT/ IN, OUT
C
        IN = 1
        OUT = 1
        WRITE(OUT, 101)
        X = 2
        CALL NEWTON(X)
        WRITE(OUT, 102) X
        STOP
101     FORMAT(/' Newton' 's method')
102     FORMAT(/' The square root of 2 is ', F8.5)
        END
        FUNCTION FUNC(X, FX, DFX)
C
C -- The square root of 2.
C
        REAL X, FX, DFX
C
        FX = X * X - 2
        DFX = 2 * X
        RETURN
        END
        SUBROUTINE NEWTON(X)
C
C -- The solution of f(x) = 0 by Newton's method.
C
        INTEGER IN, OUT
        REAL X, FX, DFX, X1, DX, TOL
        COMMON /INOUT/ IN, OUT
        DATA TOL/1.0E-6/
C
        WRITE(OUT, 101)
10        X1 = X
          CALL FUNC(X, FX, DFX)
          DX = FX / DFX
          X = X1 - DX
          WRITE(OUT, 102) X1, FX, DFX
        IF (ABS(DX) .GT. ABS(X * TOL)) GOTO 10
        RETURN
101     FORMAT('      X            FX         DFX')
102     FORMAT(1X, 0PF8.5, 3X, 1P2E12.4)
        END
```

Figure 8.6: Newton's Method, Version One

Tolerance

There are a few other matters we should now consider. Successive approximations are provided by a loop in subroutine NEWTON. This loop continues until two successive values are within the desired tolerance. We are not interested in whether the difference (DX) between two successive approximations has a negative or a positive value. We are only concerned with the magnitude. For this reason, we must be careful to take the absolute value of the comparison.

Furthermore, we are not interested in the actual difference, but only the relative difference. Suppose, for example, that we want our answers to be accurate to one part in a million, that is, one part in 10^6. If a particular solution has a value of unity, then two successive values must be closer than 10^{-6}. If, however, the solution itself has a value of 10^{-6}, then two successive values must differ by no more than 10^{-12}. We therefore choose a relative criterion rather than an absolute criterion for termination of the iteration process.

Running the Program

Type up the program shown in Figure 8.6 and try it out. The first approximation to the square root of 2 is chosen to be 2. When the program is executed, it should produce the square root of 2 after several iterations. The results will resemble Figure 8.7.

```
   Newton's method
      X               FX           DFX
   2.00000        2.0000E+00    4.0000E+00
   1.50000        2.5000E-01    3.0000E+00
   1.41667        6.9444E-03    2.8333E+00
   1.41422        5.9605E-06    2.8284E+00
   1.41421       -1.1921E-07    2.8284E+00

   The square root of 2 is  1.41421
```

Figure 8.7: Output: The Positive Root of f(x) = x² − 2

In the following sections we will make several small changes to refine this program. The first change will allow us to input different first-approximation values for the root. This facility is important for studying equations that have more than one root.

Generalizing Procedure Calls

Another matter we should consider is the relationship of the Newton's method subroutine to the subroutine it calls (FUNC) for evaluation of the function and its derivative. The Newton's method subroutine gives directions for carrying out the operation described by Equation 4. It is independent of the actual function it is operating on. Consequently, the subroutine that calculates the function and its derivative should be an entirely separate entity.

Ideally, subroutine FUNC, called by the Newton's method subroutine, should be a dummy parameter. The actual subroutine name should be passed to subroutine NEWTON as a parameter during execution. With this arrangement, the Newton's method subroutine could be directed to solve one equation at one point and another equation at another point. This approach is easily accomplished in FORTRAN, and we will take advantage of this feature in a later version of this program. We must be careful, however, to declare the actual function name in an EXTERNAL statement at the beginning of the program that calls NEWTON.

Adding User Input for the First Approximation

When the first version of Newton's method is working properly, you can begin to add new features. Make a duplicate copy of the first version. Alter the main program so that it looks like Figure 8.8. Notice that the subroutine called by NEWTON has been renamed SQR2 and is declared in an EXTERNAL statement. The name FUNC is now a dummy parameter.

```
C          PROGRAM NEWT2
C
           INTEGER IN, OUT
           REAL X
           EXTERNAL SQR2
           COMMON /INOUT/ IN, OUT
C
           IN = 1
           OUT = 1
           WRITE(OUT, 101)
10         WRITE(OUT, 103)
           READ(IN, 104) X
           IF (X .LT. -19.0) GOTO 99
           CALL NEWTON(X, SQR2)
           WRITE(OUT, 102) X
           GOTO 10
99         STOP
101        FORMAT(/' Newton' 's method')
102        FORMAT(/' The square root of 2 is ', F8.5)
103        FORMAT(' First guess? ')
104        FORMAT(E10.0)
           END
           FUNCTION SQR2(X, FX, DFX)
C
C  -- The square root of 2.
C
           REAL X, FX, DFX
C
           FX = X * X - 2
           DFX = 2 * X
           RETURN
```

Figure 8.8: The Main Program for Version Two

```
            END
            SUBROUTINE NEWTON(X, FUNC)
      C
      C  -- The solution of f(x) = 0 by Newton's method.
      C
            INTEGER IN, OUT
            REAL X, FX, DFX, X1, DX, TOL
            COMMON /INOUT/ IN, OUT
            DATA TOL/1.0E-6/
      C
            WRITE(OUT, 101)
      10      X1 = X
            CALL FUNC(X, FX, DFX)
            DX = FX / DFX
            X = X1 - DX
            WRITE(OUT, 102) X1, FX, DFX
            IF (ABS(DX) .GT. ABS(X * TOL)) GOTO 10
            RETURN
      101   FORMAT('      X              FX        DFX')
      102   FORMAT(1X, 0PF13.5, 3X, 1P2E12.4)
            END
```

Figure 8.8: The Main Program for Version Two (cont.)

Run the new version to try it out. For the first version, the value of 2 was encoded into the program as the first guess to Newton's method. The second version is more sophisticated than the first. The user is asked to input the first guess. The successive approximations, along with the function and its derivative, are displayed as before. At the conclusion of the task, the program begins again and the user is asked to input another first approximation.

Running the Program to Find the Second Root

Start with the value of 2; the results should be the same as they were for the first version. Then, for the second cycle, try the value of 1. This first approximation is on the other side of the root, but the square root of 2 should again be found in relatively few steps. Try the value of -2 for the third cycle. Notice that the process converges on a different root this time. There are, of course, two solutions to the equation:

$$x^2 - 2 = 0$$

We found the negative root this time by giving a negative first approximation.

Investigate what happens when the first guess is near the midpoint of the two roots. Try a first guess of 0.0001. Now, the process takes quite a few steps to produce the answer. Finally, try a first guess of zero. The curve

$$y = f(x)$$

has zero slope at this point. Consequently, one of two things can happen.

Either a floating-point divide error can occur or the program can loop in-
definitely. This problem will be corrected in the next version.

A Test for Zero Slope

When the derivative, or slope, of our function is zero, the final term in
Equation 4 becomes infinite. We are looking for the point where the slope
crosses the x-axis. But it can be seen from Figure 8.9 that the tangent to the
curve at $x = 0$ is parallel to the x-axis. Consequently, the two lines do not
intersect.

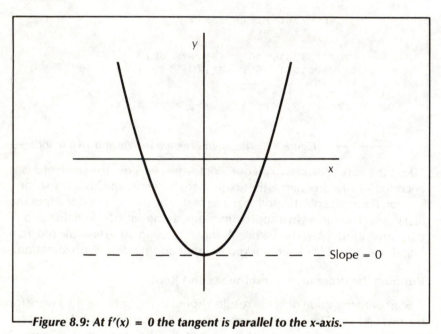

Figure 8.9: At f'(x) = 0 the tangent is parallel to the x-axis.

We will now add some instructions for testing the slope to our Newton's
method program. One way to do this is to define a small number such as:

DATA SMALL/1E−15/

The slope can then be tested with the statement:

IF (ABS(DFX) .LT. SMALL) GOTO 99

One problem with this approach, however, is that the value assigned to
SMALL must be consistent with the particular version of FORTRAN being
used. That is, the value of SMALL may have to be chosen carefully.

The check for zero slope that we will actually use is more straightforward:

IF (DFX .EQ. 0.0) GOTO 99

If the slope is found to be zero, an error message is printed and the error flag is set. Otherwise, the process continues normally. Notice that the error flag is available to the main program because it is the third parameter of the subroutine. If the error flag is not set, then the main program will print the value of X. Compile the new version shown in Figure 8.10 and try it out.

```
C       PROGRAM NEWT3
C
        LOGICAL ERROR
        INTEGER IN, OUT
        REAL X
        EXTERNAL SQR2
        COMMON /INOUT/ IN, OUT
C
        IN = 1
        OUT = 1
        WRITE(OUT, 101)
10      WRITE(OUT, 103)
        READ(IN, 104) X
        IF (X .LT. -19.0) GOTO 99
        CALL NEWTON(X, SQR2, ERROR)
        IF (.NOT. ERROR) WRITE(OUT, 102) X
        GOTO 10
99      STOP
101     FORMAT(/' Newton' 's method')
102     FORMAT(/' The square root of 2 is ', F8.5)
103     FORMAT(' First guess? ')
104     FORMAT(E10.0)
        END
        FUNCTION SQR2(X, FX, DFX)
C
C -- The square root of 2.
C
        REAL X, FX, DFX
C
        FX = X * X - 2
        DFX = 2 * X
        RETURN
        END
        SUBROUTINE NEWTON(X, FUNC, ERROR)
C
C -- The solution of f(x) = 0 by Newton's method.
C
        LOGICAL ERROR
        INTEGER IN, OUT
        REAL X, FX, DFX, X1, DX, TOL
        COMMON /INOUT/ IN, OUT
        DATA TOL/1.0E-6/
C
        ERROR = .FALSE.
        WRITE(OUT, 101)
```

Figure 8.10: A test for zero slope is added.

```
10        X1 = X
          CALL FUNC(X, FX, DFX)
          IF (DFX .EQ. 0.0) GOTO 99
          DX = FX / DFX
          X = X1 - DX
          WRITE(OUT, 102) X1, FX, DFX
        IF (ABS(DX) .GT. ABS(X * TOL)) GOTO 10
        RETURN
99        ERROR = .TRUE.
          WRITE(OUT, 999)
          RETURN
101       FORMAT('              X              FX          DFX')
102       FORMAT(1X, 0PF13.5, 3X, 1P2E12.4)
999       FORMAT(' ERROR--slope zero')
          END
```

Figure 8.10: A test for zero slope is added. (cont.)

Running the Program with the Slope Test

Give initial values of 2, 1, and −1 as before to see that the program behaves properly. Then, give a first approximation of zero. This input caused a floating-point divide check in the previous version. Now, however, your FORTRAN program should handle this input with no diffi- culty. The program should print the appropriate error message, then request another first approximation. The program can be aborted by entering a value less than −19.

Finally, our next task is to make sure the program will print an error message and terminate after an appropriate number of iterations if approx- imations do not converge on a root.

Failure to Converge

Sometimes, Newton's method will not converge on a root after a reason- able number of iterations. One possibility is that successive approximations are oscillating around a complex root, as illustrated in Figure 8.11. The first approximation, x_1, gives rise to the second value, x_2. But x_2 then produces the original value of x_1 for the third approximation. As a result, the process will never terminate.

Another possibility is that an approximation is very far from a root. This can occur even though the first guess appears to be close to a root. We can observe this behavior with our previous version of Newton's method if we run the program again and give a first approximation of 0.00001. This will produce a second value of 20,000, which is quite an overshoot. Each succes- sive approximation will now be about one-half of the previous value. A solution will eventually be obtained, but it will take more than 20 iterations before convergence.

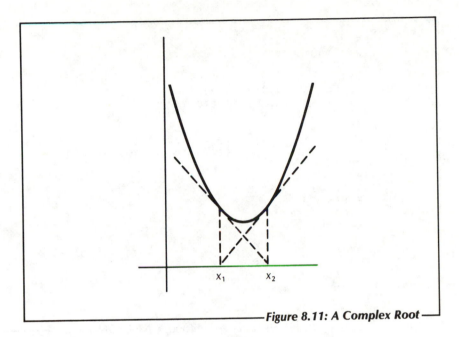

Figure 8.11: A Complex Root

We can guard against lack of convergence by adding a loop counter. We can then abort the process if convergence does not occur after, say, 20 iterations. This will take care of the case of oscillation, as well as the case of an approximation that is very far from a root.

Alter subroutine NEWTON so that it looks like Figure 8.12. Run the new version to try it out. Give an initial approximation of 500000. This first guess is so far from the root that it will take many cycles to converge. The iteration process should stop after 20 loops and the appropriate error message will be displayed. Next, try a first guess of 2 to make sure that everything is still all right.

```
        SUBROUTINE NEWTON(X, FUNC, ERROR)
C
C -- The solution of f(x) = 0 by Newton's method.
C
        LOGICAL ERROR
        INTEGER IN, OUT, MAX, I
        REAL X, FX, DFX, X1, DX, TOL
        COMMON /INOUT/ IN, OUT
        DATA TOL/1.0E-6/, MAX/20/
C
```

Figure 8.12: A loop counter is added to subroutine NEWTON.

```
          ERROR = .FALSE.
          WRITE(OUT, 101)
          DO 20 I = 1, MAX
            X1 = X
            CALL FUNC(X, FX, DFX)
            IF (DFX .EQ. 0.0) GOTO 99
            DX = FX / DFX
            X = X1 - DX
            WRITE(OUT, 102) X1, FX, DFX
          IF (ABS(DX) .LE. ABS(X * TOL)) GOTO 30
20        CONTINUE
          ERROR = .TRUE.
          WRITE(OUT, 998) MAX
30        RETURN
99        ERROR = .TRUE.
          WRITE(OUT, 999)
          RETURN
101       FORMAT('            X              FX        DFX')
102       FORMAT(1X, 0PF13.5, 3X, 1P2E12.4)
998       FORMAT(' ERROR--no convergence in', I3, ' loops')
999       FORMAT(' ERROR--slope zero')
          END
```

Figure 8.12: A loop counter is added to subroutine NEWTON. (cont.)

Let us now take advantage of the dummy parameter name FUNC we introduced earlier, so we can solve two different equations with the same Newton's method subroutine.

FORTRAN PROGRAM: SOLVING TWO DIFFERENT EQUATIONS

The advantage of declaring the subroutine FUNC as a dummy parameter to subroutine NEWTON is that we can solve more than one equation with the same Newton's method routine. Let us therefore add a routine to solve a second equation: the cube root of 2. Figure 8.13 shows the revised main program, which calls both SQR2 and CUBE. Function subprogram CUBE is also shown. Of course, subroutine SQR2 and NEWTON must also be provided. Notice that the name CUBE has been added along with SQR2 to the EXTERNAL statement in the main program.

```
C       PROGRAM NEWTC
C
C -- Square and cube root of 2 by Newton's method.
C
        LOGICAL ERROR
        INTEGER IN, OUT
        REAL X
```

Figure 8.13: Solving Two Different Equations

```
            EXTERNAL SQR2, CUBE
            COMMON /INOUT/ IN, OUT
C
            IN = 1
            OUT = 1
            WRITE(OUT, 101)
            X = 2
            CALL NEWTON(X, SQR2, ERROR)
            WRITE(OUT, 102) X
            CALL NEWTON(X, CUBE, ERROR)
            WRITE(OUT, 105) X
99          STOP
101         FORMAT(/' Newton' 's method')
102         FORMAT(/' The square root of 2 is ', F8.5)
105         FORMAT(//' The cube root of 2 is ', F8.5)
            END
            FUNCTION CUBE(X, FX, DFX)
C
C  -- The cube root of 2.
C
            REAL X, FX, DFX, X2
C
            X2 = X * X
            FX = X2 * X - 2
            DFX = 3 * X2
            RETURN
            END
```

Figure 8.13: Solving Two Different Equations (cont.)

We have developed and refined our program using simple, predictable functions. Now we are ready to use our program to find the roots of some more difficult functions.

FORTRAN PROGRAM: SOLVING OTHER EQUATIONS

Consider the nonlinear equation:

$$e^x = 4x$$

This equation, unlike our previous one, cannot be solved explicitly for x. Consequently, it is a suitable candidate for our Newton's method program. The corresponding function, and its derivative are:

$$f(x) = e^x - 4x$$

and

$$f'(x) = e^x - 4$$

Let us find the roots of this equation with our Newton's method program.

Alter the main program so that it is more general. That is, delete the lines referring to the cube root, and change FORMAT statement 102 to remove the square root reference. The name SQR2 is also changed to the more general FUNC. The new main program and subroutine FUNC are given in Figure 8.14. Notice that the variable E is introduced so that the exponent of X will not have to be calculated twice. Recall that:

$$\frac{de^x}{dx} = e^x$$

```
C       PROGRAM NEWT5
C
        LOGICAL ERROR
        INTEGER IN, OUT
        REAL X
        EXTERNAL FUNC
        COMMON /INOUT/ IN, OUT
C
        IN = 1
        OUT = 1
        WRITE(OUT, 101)
10      WRITE(OUT, 103)
        READ(IN, 104) X
        IF (X .LT. -19.0) GOTO 99
        CALL NEWTON(X, FUNC, ERROR)
        IF (.NOT. ERROR) WRITE(OUT, 102) X
        GOTO 10
99      STOP
101     FORMAT(/' Newton' 's method')
102     FORMAT(/' The answer is ', F12.5)
103     FORMAT(' First guess? ')
104     FORMAT(E10.0)
        END
        FUNCTION FUNC(X, FX, DFX)
C
C  --  The solution for f(X) = exp(X) - 4X.
C
        REAL X, FX, DFX, E
C
        E = EXP(X)
        FX = E - 4.0 * X
        DFX = E - 4.0
        RETURN
        END
```

Figure 8.14: Subroutine for Evaluating $e^x = 4x$

Run the new version and input a first guess of 4. The program should converge on a value of 2.153. This is one of the two roots. The other root can be found by giving a first guess of 0.1; it has a value of 0.357.

Because our next equation involves the SIN function, its implementation will require one of two versions, depending on the FORTRAN in use.

A Function with Many Roots

Let us explore the several roots of the equation:

$$\sin(x) = \frac{x}{10}$$

Change subroutine FUNC so that it looks like Figure 8.15.

```
        FUNCTION FUNC(X, FX, DFX)
C
C  -- The solution to f(X) = sin(X) - X/10.
C
        REAL X, FX, DFX
C
        FX = SIN(X) - 0.1 * X
        DFX = COS(X) - 0.1
        RETURN
        END
```

Figure 8.15: The Solution of Sin(x) = x / 10

Zero is one root of the new equation. As we approach this root, we will need to take the SIN of numbers that are closer and closer to zero. Unfortunately, as we discovered in Chapter 1, some FORTRAN compilers contain an error that will cause trouble under these conditions. The problem is that an incorrect value may be returned for the SIN as the argument approaches zero.

If you haven't tested your SIN function with the program given in Chapter 1, you should do so now. If the results of the test indicate a problem with your SIN, then you should use the subroutine given in Figure 8.16 rather than the one in Figure 8.15. The alternate approach is to inspect the argument of the SIN. If the number is closer to zero than a value such as 0.00001, then the value of the argument is returned. Otherwise, the built-in SIN function is called.

Run this latest version to try it out. Give a first estimate of 1.0. This should converge on the root at zero. Then, try a first guess of −1.0. This too should converge on the root at zero. There are several other roots to this equation. The table in Figure 8.17 gives several first approximations and the corresponding roots. Try out each one to verify that your Newton's method program is working properly.

```
         FUNCTION FUNC(X, FX, DFX)
C
C -- The solution to f(X) = sin(X) - X/10.
C
         REAL X, FX, DFX, SMALL
         DATA SMALL/1.0E-5/
C
         IF (ABS(X) .LT. SMALL) GOTO 10
         FX = SIN(X) - 0.1 * X
         DFX = COS(X) - 0.1
         RETURN
10       FX = X - 0.1 * X
         DFX = 1 - 0.1
         RETURN
         END
```

Figure 8.16: Alternate Implementation for sin(x) = x / 10

1st approximation	root
−1	0
1	0
4	2.852..
4.3	7.068..
4.5	0
4.7	−8.423..
5	−2.852..
6	7.068..
9	8.423..

Figure 8.17: Some Roots to f(x) = sin(x) − x/10

FORTRAN PROGRAM: THE VAPOR PRESSURE EQUATION

We are now ready to solve the vapor pressure equation that was introduced at the beginning of this chapter. We will write our function and its derivative as:

$$f(T) = A + \frac{B}{T} + C \ln T - \ln P$$

$$df(T) = \frac{-B}{T^2} + \frac{C}{T}$$

Remember that A, B, C, and P are constants. Change subroutine FUNC so it looks like Figure 8.18. Notice that the variable LOGP is defined as the logarithm of 0.01. This step avoids repeated calculations of the logarithm. With this function we will find the temperature that corresponds to a vapor pressure of 0.01 atmospheres.

```
      FUNCTION FUNC(T, FT, DFT)
C
C  -- The solution to f(T) = A + B/T + C Log(T) - Log(.01).
C
      REAL T, FT, DFT, A, B, C, LOGP
      DATA A/18.19/, B/-23180.0/, C/-0.8858/ LOGP/-4.60517/
C
      FT = A + B / T + C * ALOG(T) - LOGP
      DFT = -B / (T*T) + C / T
      RETURN
      END
```

Figure 8.18: Solution of the Vapor Pressure Equation

Run the new version to try it out. Give a first guess of 500 degrees and the program will converge to a temperature of 1416 degrees in about seven steps. Then try a first guess of 2000 degrees. The solution will converge to 1416 degrees in about six steps.

SUMMARY

Using a familiar function at first, we saw how easy it was to write, and then improve upon, a FORTRAN program implementing Newton's method for finding the roots of an equation. We then used our program to solve some more complex functions. These included exponential and trigonometric functions. Finally, we solved the equation of an actual scientific application.

EXERCISES

8-1: The degree of dissociation, x, of hydrogen sulfide gas can be described by the expression:

$$(1 - PK^2) x^3 - 3x + 2 = 0$$

where K is the equilibrium constant, and P is the total pressure in atmospheres. For a temperature of 2000 kelvins, K = 0.608. Use Newton's method to find the degree of dissociation when the pressure is one atmosphere. Since x only has meaning over the range 0–1, begin with an approximation of 0.5. Explore the consequences of initial guesses of zero, of unity, and of 1.2.

Answer: x = 0.758

8-2: Van der Waal's equation of state is:

$$(P + A/V^2)(V - B) = RT$$

where P is the pressure, V is the molar volume, T is the temperature and R is the gas constant. Coefficient A is a measure of the bonding force and coefficient B is a volume correction. When the pressure is stated in atmospheres, the volume in liters and the temperature in kelvins, R has a value of 0.082 liter-atm/deg mole. Coefficient A is in units of pressure times molar volume squared and B has units of molar volume. The coefficients A and B have been extensively tabulated for common gases. For example, the values for toluene are:

$$A = 24.06 \text{ liter}^2 \text{ atm/mole}^2$$

and

$$B = 0.1463 \text{ liter/mole}$$

Use Newton's method to solve the Van der Waals equation. Find the molar volume of toluene at the boiling point of 110°C (383° K) and a pressure of 1 atm. Write a subroutine VANDER(V, F, DF). Define the values for A, B, P, R, and T in a DATA statement of this subroutine. Then define F and DF:

F = V − R*T/P − B + A/(P*V) − A*B/(V*V)
DF = 1.0 − A/(P*V*V) + A*B/(V*V*V)

Change the main program so that the program calculates the first approximation from the ideal gas law: V = RT/P. Remove the GOTO 10 statement in the main program. Run the program to find the Van der Waals volume.

Answer: 30.8 liters

8-3: Use Newton's method to solve the Van der Waals equation as in the previous problem, but input the temperature and pressure from the keyboard. Print both the ideal gas volume and the Van der Waals volume.

9

Numerical Integration

IN THIS CHAPTER, we will develop three different methods for carrying out numerical integration; that is, for determining the area underneath a curve between two given values of the independent variable. After writing FORTRAN programs for each of the methods, we will judge how efficiently each method computes the area. Each of the methods uses progressively smaller "panels" of measurable area to divide the area beneath the curve. The summation of the areas of these panels then provides an approximation of the total area.

In the trapezoidal rule method, these panels are topped by straight-line secants to the curve, whereas Simpson's method tops each panel with a parabolic curve of its own. Both of these methods can be improved with an abbreviated Taylor series expansion to compute the error—a method called *end correction*.

The third integration method we will study, the Romberg method, is the most complex; it constructs a matrix of interpolations to arrive at the total area. One function we will try on both the Simpson and Romberg methods is related to the normal distribution function, to which we will be returning in Chapter 11. Finally, we will explore a method for integrating a function that approaches infinity at one of its limits.

Let us begin with a brief description of the definite integral and the methods of evaluating it.

THE DEFINITE INTEGRAL

The evaluation of the *definite integral:*

$$\int_a^b f(x)dx = F(b) - F(a)$$

where $F'(x) = dF(x)/dx = f(x)$, can be interpreted as the area under the curve of the function $f(x)$ from the limit a to the limit b, as illustrated in Figure 9.1.

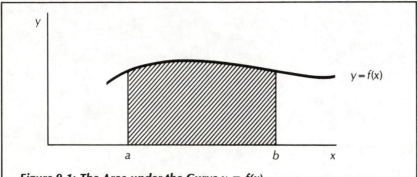

—Figure 9.1: The Area under the Curve $y = f(x)$—

The actual integration can be straightforward for certain functions, but very difficult for others. For example, the power series:

$$1 + 2x + 3x^2$$

can be integrated term by term and then evaluated between the limits of a and b to give:

$$b + b^2 + b^3 - a - a^2 - a^3$$

More complicated functions can sometimes be integrated by reference to integration tables found in math handbooks. Other techniques, such as *integration by parts,* can sometimes be used. Another possibility is to replace the original function with an infinite series or an *asymptotic expansion.* The integration is then carried out on the replacement function.

Sometimes the integrand is not a proper function at all, but simply a collection of experimental data. In such a case, it may be possible to fit the experimental data to a function that can be integrated. Nevertheless, there will be times when it will be impossible or very difficult to evaluate the integrand analytically. But if the limits are known, then an approximation method may provide an acceptable solution. This approach is known as *numerical integration.*

Several different methods are commonly used for numerical integration. These typically involve the substitution of an easily integrated function for the original function. The new function may be a polynomial such as a straight line or parabola, or it may consist of transcendental functions such as sines and cosines. The accuracy of the resulting calculation depends on how well the substituted function approximates the original function.

In the next section we will describe both the general method of function substitution, and a specific, simple implementation of the method—the use of straight-line functions.

THE TRAPEZOIDAL RULE

One of the simplest methods of numerical integration is known as the *trapezoidal rule*. In this method, the original function is approximated by a set of straight lines. The region to be integrated is divided into uniformly spaced sections or panels. The panel width, Δx, is

$$\Delta x = \frac{b - a}{n}$$

where n is the number of panels and a and b are the integration limits. If the entire integral is fitted with a single straight line, that is, if there is only one panel (as illustrated in Figure 9.2) then the calculated area is:

$$\frac{(b - a)[f(a) + f(b)]}{2}$$

In this formula, $f(a)$ is the value of the function at the left limit and $f(b)$ is the value at the right limit.

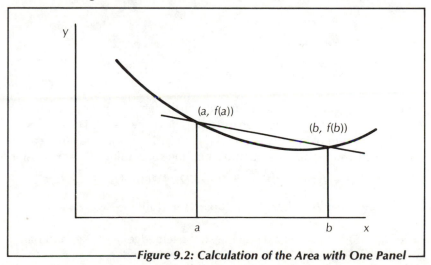

Figure 9.2: Calculation of the Area with One Panel

For the more general case, the area is divided into n panels. An interior panel is bounded on the left by a vertical line at x_i, and on the right by a vertical line at $x_i + \Delta x$. The lower edge of the panel is marked by the x-axis. The original curve at the top of the panel is replaced by a straight line that in general will not have a zero slope (that is, it will not be parallel to the x-axis). The resulting panel thus has the shape of a trapezoid, giving rise to the name of the method.

The panels can be numbered from 1 to n, but actually we are interested in evaluating the function at the left and right edges of each panel. There are $n - 1$ interior edges for the n panels, in addition to the right and left boundaries of the integral. Consequently, there will be $n + 1$ edges, which can be numbered from 0 through n.

The area of the first panel is:

$$\frac{[f(0) + f(1)]\Delta x}{2}$$

or

$$\frac{[f(a) + f(1)]\Delta x}{2}$$

and the area of the last panel is:

$$\frac{[f(n-1) + f(n)]\Delta x}{2}$$

or

$$\frac{[f(n-1) + f(b)]\Delta x}{2}$$

The area of the ith panel is:

$$\frac{[f(i-1) + f(i)]\Delta x}{2}$$

where $f(i-1)$ is the value of the function on the left side of the ith panel and $f(i)$ is the value of the function on the right side of the panel. The desired integral is the sum of the areas of all the panels. Thus, the total area can be calculated by summing the areas of all the panels according to the expression:

$$\{[f(a) + f(1)] + [f(1) + f(2)] + [f(2) + f(3)] + \ldots$$
$$+ [f(n-2) + f(n-1)] + [f(n-1) + f(b)]\} \ \frac{\Delta x}{2}$$

The right edge of the first panel is also the left edge of the second panel, and the left edge of the last panel is also the right edge of the next to last panel. The edges of all the other panels are also common to two panels. The formula for integration by the trapezoidal rule can therefore be simplified as:

$$[f(a) + 2f(1) + 2f(2) + \ldots$$
$$+ 2f(n-2) + 2f(n-1) + f(b)] \ \frac{\Delta x}{2}$$

Our first computer program for this method will allow us to experiment with the number of panels we use to partition the total area.

FORTRAN PROGRAM: THE TRAPEZOIDAL RULE WITH USER INPUT FOR THE NUMBER OF PANELS

The area calculated from the trapezoidal method more closely approaches the actual value as the number of panels becomes larger and larger. This can be demonstrated with the program given in Figure 9.3. Type up the program and run it. The main program will ask for the number of sections into which the area is to be divided. The resulting calculated value will then be based on the given number of panels.

```
C       PROGRAM TRAP1
C
        INTEGER IN, OUT, PIECES
        REAL SUM, UPPER, LOWER
        COMMON /INOUT/ IN, OUT
        DATA LOWER/1.0/, UPPER/9.0/
C
        IN = 1
        OUT = 1

        WRITE(OUT, 101)
10      WRITE(OUT, 102)
        READ(IN, 103) PIECES
        IF (PIECES .LT. 0) GOTO 99
        CALL TRAPEZ(LOWER, UPPER, PIECES, SUM)
        WRITE(OUT, 104) SUM
        GOTO 10
99      STOP
101     FORMAT(/' Trapezoidal integration')
102     FORMAT(' How many pieces? ')
103     FORMAT(I3)
104     FORMAT(' Area =', F10.5)
        END
        SUBROUTINE TRAPEZ(LOWER, UPPER, PIECES, SUM)
C
C -- Numerical integration by the trapezoid method.
C
```

Figure 9.3: Integration by the Trapezoidal Rule

```
         INTEGER IN, OUT, PIECES, I
         REAL X, DELTA, ENDSUM, MIDSUM, LOWER, UPPER, SUM
C
C -- f(X) = 1 / X, be careful of X = 0.
C
         F(X) = 1.0 / X
C
         DELTA = (UPPER - LOWER) / PIECES
         ENDSUM = F(LOWER) + F(UPPER)
         MIDSUM = 0.0
         DO 10 I = 1, PIECES
            X = LOWER + I * DELTA
            MIDSUM = MIDSUM + F(X)
10       CONTINUE
         SUM = (ENDSUM + 2.0*MIDSUM) * DELTA * 0.5
         RETURN
         END
```

Figure 9.3: Integration by the Trapezoidal Rule (cont.)

The algorithm for integration by the trapezoidal rule is contained in subroutine TRAPEZ. The function to be integrated:

$$\int_1^9 \frac{dx}{x}$$

is defined as a statement function, F, at the beginning of TRAPEZ. Since this function can be readily integrated, we can compare the exact value of the integral to the value calculated by the trapezoidal method. The value of this integral is the natural logarithm of 9, or 2.197225.

We will now look at a more sophisticated version of this program.

FORTRAN PROGRAM: AN IMPROVED TRAPEZOIDAL RULE

We can improve our trapezoidal program in two ways. First, we can change subroutine TRAPEZ so that it will automatically divide the original area into more and more pieces. Second, we can avoid much calculation at each step by using the results from the previous step. A third improvement moves the statement function F to the beginning of the main program. The function F in subroutine TRAPEZ then becomes a dummy variable.

Alter the first trapezoidal program so that it looks like the one given in Figure 9.4 and then execute it.

```
C          PROGRAM TRAP2
C
           INTEGER IN, OUT
           REAL SUM, UPPER, LOWER, TOL
           COMMON /INOUT/ IN, OUT
           DATA LOWER/1.0/, UPPER/9.0/, TOL/1.0E-5/
C
C -- f(X) = 1 / X, be careful of X = 0.
C
           F(X) = 1.0 / X
C
           IN = 1
           OUT = 1

           WRITE(OUT, 101)
           CALL TRAPEZ(LOWER, UPPER, TOL, SUM, F)
           WRITE(OUT, 104) SUM
           STOP
101        FORMAT(/' Trapezoidal integration')
104        FORMAT(/' Area =', F10.5/)
           END
           SUBROUTINE TRAPEZ(LOWER, UPPER, TOL, SUM, F)
C
C -- Numerical integration by the trapezoid method.
C
           INTEGER IN, OUT, PIECES, I, P2
           REAL X, DELTA, LOWER, UPPER, SUM, TOL
           REAL ENDSUM, MIDSUM, SUM1
           COMMON /INOUT/ IN, OUT
C
           PIECES  = 1
           DELTA = (UPPER - LOWER) / PIECES
           ENDSUM = F(LOWER) + F(UPPER)
           SUM = ENDSUM * DELTA / 2.0
           WRITE(OUT, 101) SUM
           MIDSUM = 0.0
5          PIECES = PIECES * 2
             P2 = PIECES / 2
             SUM1 = SUM
             DELTA = (UPPER - LOWER) / PIECES
             DO 10 I = 1, P2
               X = LOWER + DELTA *(2 * I - 1)
               MIDSUM = MIDSUM + F(X)
10           CONTINUE
             SUM = (ENDSUM + 2.0*MIDSUM) * DELTA * 0.5
             WRITE(OUT, 102) PIECES, SUM
           IF (ABS(SUM - SUM1) .GT. ABS(TOL * SUM)) GOTO 5
           RETURN
101        FORMAT(/'        1', F9.5)
102        FORMAT(1X, I7, F9.5)
           END
```

Figure 9.4: An Improved Trapezoidal Method

Running the Program

For the first calculation, the entire area is taken as a single panel. The number of panels is then doubled to 2 and the new area is calculated. The process is continued with a doubling of the number of panels at each step. As the number of panels increases, so does the accuracy of the result and, of course, the length of the computation time. The number of panels and the corresponding calculated areas are displayed at each step. The output should look like Figure 9.5.

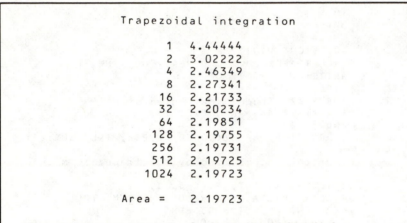

```
          Trapezoidal integration

           1    4.44444
           2    3.02222
           4    2.46349
           8    2.27341
          16    2.21733
          32    2.20234
          64    2.19851
         128    2.19755
         256    2.19731
         512    2.19725
        1024    2.19723

     Area  =    2.19723
```

Figure 9.5: Output: Integration by the Trapezoidal Method
(The number of panels and the corresponding areas are given.)

It can be seen from Figure 9.5 that the calculated value more closely approaches the correct value as the number of panels is increased. The process terminates when two successive values are within the desired tolerance. The tolerance (TOL) is set to a value of 10^{-5} in the main program and is passed as a parameter to subroutine TRAPEZ. You may want to change this value to correspond to the precision of your FORTRAN. You may also want to define the value of TOL in the subroutine rather than the main program. This simplifies the calling sequence at the expense of versatility. In this case, however, TOL should not be a parameter to the subroutine. Another possible alteration is the removal of the intermediate print statements from subroutine TRAPEZ.

With this version, the number of panels at each step is twice the number used for the previous step. Notice that it is not necessary to recalculate all the panel heights at each step. Suppose, for example, that four panels are used for a particular step. The next step will then use eight panels. Since all the panel heights from the previous step are common to the present step, it is faster to save the sum from each step and use it for the next step, rather than recalculate all the heights. Only the heights midway between the previous points need to be computed. These are then added to the previous sum of the interior values. The new sum is multiplied by 2 and added to the values at each end of the function. This result is then multiplied by one-half the panel width to obtain the area.

Although this version appears to be more efficient than the first version, it suffers from the same errors. We are trying to fit a general curve with a set of straight lines. If the original curve is not a straight line, many steps may be needed for convergence.

In our final version of the trapezoidal rule method we will include a calculation of end correction in our program.

End Correction

We can further improve the trapezoidal method by including correction factors obtained from a Taylor series expansion. The error terms contain second derivatives of the function. Fortunately, however, simplified correction terms can be used instead. The terms for the interior points vanish, leaving only the terms at the upper and lower limits of the function. Thus, the major error-correction term requires a derivative of the function at the two limits, x_0 and x_n. The additional quantity to be included in the sum is:

$$\frac{[f'(b) - f'(a)](\Delta x)^2}{12}$$

where $f'(b)$ is the value of the slope at b and $f'(a)$ is the value of the slope at a. The resulting value is subtracted from the regular trapezoidal sum. The method is known as integration by the trapezoidal rule with end correction.

FORTRAN PROGRAM: TRAPEZOIDAL RULE WITH END CORRECTION

Make a copy of the previous program and alter it to look like Figure 9.6. Notice that a second statement function, DF, has been added just after the function F at the beginning of the main program. DF is also added as a parameter to the subroutine call.

```
C       PROGRAM TRAP3
C
        INTEGER IN, OUT
        REAL SUM, UPPER, LOWER, TOL
        COMMON /INOUT/ IN, OUT
        DATA LOWER/1.0/, UPPER/9.0/, TOL/1.0E-5/
```

Figure 9.6: Trapezoidal Rule with End Correction

```
C
C -- f(X) = 1 / X, be careful of X = 0.
C
      F(X) = 1.0 / X
      DF(X) = -1.0/(X * X)
C
      IN = 1
      OUT = 1
      WRITE(OUT, 101)
      CALL TRAPEZ(LOWER, UPPER, TOL, SUM, F, DF)
      WRITE(OUT, 104) SUM
      STOP
101   FORMAT(/' Trapezoidal integration with end correction')
104   FORMAT(/' Area =', F10.5/)
      END
      SUBROUTINE TRAPEZ(LOWER, UPPER, TOL, SUM, F, DF)
C
C -- Numerical integration by the trapezoid method.
C
      INTEGER IN, OUT, PIECES, I, P2
      REAL X, DELTA, LOWER, UPPER, SUM, TOL
      REAL ENDSUM, MIDSUM, SUM1, ENDCOR
      COMMON /INOUT/ IN, OUT
C
      PIECES  = 1
      DELTA = (UPPER - LOWER) / PIECES
      ENDSUM = F(LOWER) + F(UPPER)
      ENDCOR = (DF(UPPER) - DF(LOWER)) / 12.0
      SUM = ENDSUM * DELTA / 2.0
      WRITE(OUT, 101) SUM
      MIDSUM = 0.0
5     PIECES = PIECES * 2
        P2 = PIECES / 2
        SUM1 = SUM
        DELTA = (UPPER - LOWER) / PIECES
        DO 10 I = 1, P2
          X = LOWER + DELTA *(2 * I - 1)
          MIDSUM = MIDSUM + F(X)
10      CONTINUE
        SUM = (ENDSUM + 2.0*MIDSUM) * DELTA * 0.5
     *    - DELTA * DELTA * ENDCOR
        WRITE(OUT, 102) PIECES, SUM
      IF (ABS(SUM - SUM1) .GT. ABS(TOL * SUM)) GOTO 5
      RETURN
101   FORMAT(/'        1', F9.5)
102   FORMAT(1X, I7, F9.5)
      END
```

Figure 9.6: Trapezoidal Rule with End Correction (cont.)

Running the Program

Run the program and compare the results to Figure 9.7. Notice how much faster the process converges in this example. Also notice that during the iteration process, the sum passed the correct value, then reversed its direction. It should be realized, however, that it will not always be possible to use the end correction technique with the trapezoidal method. This will be the case if the slope at one limit or the other is infinite or becomes very large. (We will discuss this case at the end of this chapter.) Also, if the function has a zero slope at the limits, the end-correction term will be zero.

```
Trapezoidal Integration with End Correction
          1   4.44444
          2   1.70535
          4   2.13427
          8   2.19111
         16   2.19675
         32   2.19719
         64   2.19722
        128   2.19723

Area =    2.19723
```

Figure 9.7: Output: Trapezoidal Integration with End Correction

The trapezoidal rule method replaces the function with a first-order polynomial (a straight line). We can usually expect a second-order polynomial (a parabola) to be a better substitute; this is the idea of the method we will implement in our next program. We will examine three different versions of this program—first with the same function we ran on the trapezoidal rule program, then with an exponential and finally with a sine function.

FORTRAN PROGRAM: SIMPSON'S INTEGRATION METHOD

Simpson's method of numerical integration is similar to the trapezoidal method except that the original curve is replaced by a set of parabolas rather than straight lines. Since a parabola can be defined by a minimum of three points, we must divide the original area into an even number of panels. Each parabola is fitted to the tops of two adjacent panels. In general, we expect parabolas to produce a better fit than straight lines. That is, we should need fewer panels to obtain satisfactory convergence. The formula is:

$$(f_0 + f_n + 4 \sum_{\substack{j=1 \\ j \text{ odd}}}^{n-1} f_j + 2 \sum_{\substack{j=2 \\ j \text{ even}}}^{n-2} f_j) \frac{\Delta x}{3}$$

First Run of the Simpson's Rule Program

Make a copy of the trapezoidal program shown in Figure 9.4 and alter it to look like Figure 9.8. Execute the new program and compare the results to Figure 9.9.

```
C       PROGRAM SIMP1
C -- Numerical integration by Simpson's rule.
C
        INTEGER IN, OUT
        REAL SUM, UPPER, LOWER, TOL
        COMMON /INOUT/ IN, OUT
        DATA LOWER/1.0/, UPPER/9.0/, TOL/1.0E-5/
C
C -- f(X) = 1 / X, be careful of X = 0
C
        F(X) = 1.0 / X
C
        IN = 1
        OUT = 1

        WRITE(OUT, 101)
        CALL SIMPS(LOWER, UPPER, TOL, SUM, F)
        WRITE(OUT, 104) SUM
        STOP
101     FORMAT(/' Simpson''s rule integration'/)
104     FORMAT(/' Area =', F10.5/)
        END
        SUBROUTINE SIMPS(LOWER, UPPER, TOL, SUM, F)
C
C -- Numerical integration by Simpson's rule.
C
        INTEGER IN, OUT, PIECES, I, P2
        REAL X, DELTA, LOWER, UPPER, SUM, TOL
        REAL ENDSUM, ODDSUM, SUM1, EVSUM
        COMMON /INOUT/ IN, OUT
C
        PIECES  = 2
        DELTA = (UPPER - LOWER) / PIECES
        ODDSUM = F(LOWER + DELTA)
        EVSUM = 0.0
        ENDSUM = F(LOWER) + F(UPPER)
        SUM = (ENDSUM + 4 * ODDSUM) * DELTA / 3.0
        WRITE(OUT, 101) PIECES, SUM
5       PIECES = PIECES * 2
          P2 = PIECES / 2
          SUM1 = SUM
          DELTA = (UPPER - LOWER) / PIECES
          EVSUM = EVSUM + ODDSUM
          ODDSUM = 0.0
          DO 10 I = 1, P2
```

Figure 9.8: Numerical Integration by Simpson's Rule

```
          X = LOWER + DELTA * (2 * I - 1)
          ODDSUM = ODDSUM + F(X)
   10     CONTINUE
          SUM = (ENDSUM + 4.0 * ODDSUM + 2.0 * EVSUM)
   *        * DELTA / 3.0
          WRITE(OUT, 101) PIECES, SUM
          IF (ABS(SUM - SUM1) .GT. ABS(TOL * SUM)) GOTO 5
          RETURN
   101    FORMAT(1X, I7, F9.5)
          END
```

Figure 9.8: Numerical Integration by Simpson's Rule (cont.)

```
          Simpson's rule integration

              2    2.54815
              4    2.27725
              8    2.21005
             16    2.19864
             32    2.19734
             64    2.19723
            128    2.19723


          Area =   2.19723
```

Figure 9.9: Output: Integration by Simpson's Rule
(The number of panels and the corresponding areas are given.)

Compare Figure 9.9 with Figures 9.5 and 9.7. At each point, the value of the integral obtained by Simpson's rule is closer to the correct answer than is the value obtained from the straight trapezoidal approach for the same number of panels. As a result, fewer operations are needed and so convergence occurs more quickly. On the other hand, the convergence is about the same as the trapezoidal method with end correction.

At each step of Simpson's method, the function is evaluated only at the odd positions. The sum for the even positions is obtained from the sum of all the previous interior positions, both even and odd.

Second Run of the Simpson Program—An Exponential Function

Let us consider a second example of integration by the Simpson method. The following function is related to the normal distribution described in Chapter 2. It is also related to the error function to be described in Chapter 11. Now we will integrate the curve from zero to a value of unity.

Make a copy of the previous program and alter it so that it will find the value of the integral:

$$\int_0^1 e^{-x^2} dx$$

Change function F(X) at the beginning of the main program so that it looks like:

F(X) = EXP(−X∗X)

Then change the integration limits so that they read:

DATA LOWER/0.0/, UPPER/1.0/

Run the new version and compare the output to Figure 9.10. Notice how rapidly the process converges for this function.

```
           Simpson's rule integration

                2        .74718
                4        .74686
                8        .74683
               16        .74682

         Area  =      .74682
```

Figure 9.10: Output: Integration of e^{-x^2} by Simpson's Rule

Third Run of the Simpson Program—A Periodic Function

All the functions we have integrated so far exhibit curvature in the same direction throughout the interval. Let us now consider the periodic function:

$$\sin^2 x$$

over the interval of 0 to 4π. This function is always positive; consequently, the integral over any region should produce a positive number. But this function has a zero value at both limits, as well as at the midpoint of the interval and at the quarter points. Consequently, the first area calculated by either the trapezoidal method or the Simpson method will give a zero result. The next approximations, however, will give positive results with Simpson's method.

Make a copy of the previous program to solve this function over the limit of 0 to 4π. Alter function F(X) to look like:

F(X) = SIN(X)∗∗2

Define π and the integration limits as:

    DATA LOWER/0.0/, PI/3.14159/

    . . .

    UPPER = 4 * PI

There is another factor that must be considered in this example. Since the function has a value of zero at several strategic locations, the calculated area for two panels and for four panels gives a result of zero. Consequently, you may have to begin the integration process in subroutine SIMPS with four panels rather than two. Compile the program and run it. Compare the output to Figure 9.11.

```
          Simpson's  rule  integration


                2        .00000
                4        .00000
                8       8.37757
               16       6.28319
               32       6.28319


        Area  =       6.28319
```

Figure 9.11: Output: The Integral of sin^2 x over 0 to 4π

We will now derive an end-correction formula for Simpson's method.

FORTRAN PROGRAM: THE SIMPSON METHOD WITH END CORRECTION

Error correction can be applied to the Simpson method, as it was to the trapezoidal rule. The principal term contains the fourth derivative of the function. However, it is sometimes better to use an approximation requiring only the first derivative. As with the trapezoidal correction, the derivative is only needed at the end points. The Simpson's rule formula with end correction now looks like:

$$\{7(f_0 + f_n) + 14 \sum_{\substack{j=2 \\ j\ even}}^{n-2} f_j + 16 \sum_{\substack{j=1 \\ j\ odd}}^{n-1} f_j + \Delta x[f'(a) - f'(b)]\} \frac{\Delta x}{15}$$

Running the Program with End Correction

Make a copy of the Simpson integration program shown in Figure 9.8. Alter it to look like Figure 9.12. Some of the changes are similar to those for the trapezoidal method with end correction. It is necessary to add a statement function that calculates the derivative (DF(X)). DF is also added to the parameter list of subroutine SIMPS.

```
C         PROGRAM SIMP2
C
C --  Numerical integration by Simpson's rule.
C
          INTEGER IN, OUT
          REAL SUM, UPPER, LOWER, TOL
          COMMON /INOUT/ IN, OUT
          DATA LOWER/1.0/, UPPER/9.0/, TOL/1.0E-5/
C
C --  f(X) = 1 / X, be careful of X = 0.
C
          F(X) = 1.0 / X
          DF(X) = -1.0 / (X * X)
C
          IN = 1
          OUT = 1
          WRITE(OUT, 101)
          CALL SIMPS(LOWER, UPPER, TOL, SUM, F, DF)
          WRITE(OUT, 104) SUM
          STOP
101       FORMAT(/' Simpson''s rule integration'/)
104       FORMAT(/' Area =', F10.5/)
          END
          SUBROUTINE SIMPS(LOWER, UPPER, TOL, SUM, F, DF)
C
C --  Numerical integration by Simpson's rule
C --  with end correction.
C
          INTEGER IN, OUT, PIECES, I, P2
          REAL X, DELTA, LOWER, UPPER, SUM, TOL
          REAL ENDSUM, ODDSUM, SUM1, EVSUM, ENDCOR
          COMMON /INOUT/ IN, OUT
C
          PIECES = 2
          DELTA = (UPPER - LOWER) / PIECES
          ODDSUM = F(LOWER + DELTA)
          EVSUM = 0.0
          ENDSUM = F(LOWER) + F(UPPER)
          ENDCOR = DF(LOWER) - DF(UPPER)
          SUM = (ENDSUM + 4 * ODDSUM) * DELTA / 3.0
          WRITE(OUT, 101) PIECES, SUM
5         PIECES = PIECES * 2
            P2 = PIECES / 2
            SUM1 = SUM
            DELTA = (UPPER - LOWER) / PIECES
            EVSUM = EVSUM + ODDSUM
            ODDSUM = 0.0
            DO 10 I = 1, P2
```

Figure 9.12: Procedure SIMPS with End Correction

```
              X = LOWER + DELTA *(2 * I - 1)
              ODDSUM = ODDSUM + F(X)
     10    CONTINUE
           SUM = (7 * ENDSUM + 16 * ODDSUM + 14 * EVSUM
     *        + ENDCOR * DELTA) * DELTA / 15.0
           WRITE(OUT, 101) PIECES, SUM
           IF (ABS(SUM - SUM1) .GT. ABS(TOL * SUM)) GOTO 5
           RETURN
    101    FORMAT(1X, I7, F9.5)
           END
```

Figure 9.12: Procedure SIMPS with End Correction (cont.)

Run the program and compare the results to Figure 9.13. Convergence now is the fastest of the four techniques we have used. As with the trapezoidal method, however, the end-correction term cannot be used if the slope approaches infinity. Furthermore, if the function has zero slope at the upper and lower limits, then the end-correction term is zero.

```
                2    2.54815
                4    2.16287
                8    2.19490
               16    2.19713
               32    2.19722
               64    2.19722

       Area =       2.19722
```

Figure 9.13: Output: Simpson's Rule with End Correction

In the next section we will consider a somewhat more complicated technique for numerical integration. We will implement this technique in a FORTRAN program and then run the program on the same three functions we examined in the section on Simpson's method. This will provide an opportunity for comparison of the two methods.

THE ROMBERG METHOD

The Simpson method, which uses a set of second-order equations, is an improvement over the trapezoidal method, in which first-order equations are used. Accordingly, we might attempt to further improve our numerical integration method by replacing the original curve with a set of cubic, or even higher-order, polynomials. But there is another approach, known as the Romberg integration method. With this technique, the area is calculated by the trapezoidal method, but the errors inherent in the trapezoidal method are accounted for by using an interpolation method.

We designate the usual sequence of the trapezoidal values by the notation t_{11}, t_{21}, t_{31}, etc. They are assigned to the first column of the two-dimensional matrix T. The first level of interpolated values is designated as t_{12}, t_{22}, t_{32}, etc. These values are placed into the second column of the T matrix. We can then interpolate between the interpolated values to produce a third column designated t_{13}, t_{23}, t_{33}, etc.

$$\begin{bmatrix} t_{11} & t_{12} & t_{13} & \cdots \\ t_{21} & t_{22} & t_{23} & \cdots \\ t_{31} & t_{32} & t_{33} & \cdots \\ \cdots & \cdots & \cdots & \cdots \end{bmatrix}$$

If we continue in this way, we will find that the interpolated values rapidly converge on the correct integral. The advantage of this method is that the function only has to be evaluated for the entries in the first column, corresponding to the regular trapezoidal-rule values. The function does not have to be evaluated to obtain the entries for the other columns. Rather, each value is obtained from a combination of the entry directly to the left and the one just below the entry to the left. For example:

$$t_{12} = \frac{4t_{21} - t_{11}}{3}$$

$$t_{22} = \frac{4t_{31} - t_{21}}{3}$$

$$t_{13} = \frac{16t_{22} - t_{12}}{15}$$

The general algorithm is:

$$T_{ij} = \frac{4^{j-1}t_{i+1,\,j-1} - t_{i,\,j-1}}{4^{j-1} - 1}$$

FORTRAN PROGRAM: INTEGRATION BY THE ROMBERG METHOD

A program that can be used to perform numerical integration by the Romberg method is given in Figure 9.14. Type up the program and execute it.

First Run of the Romberg Program

The regular trapezoidal values are printed at each step, as before. In addition, the interpolated values at the right side of the matrix are also shown. The results should look like Figure 9.15.

```
C       PROGRAM ROMB1
C
C -- Numerical integration by the Romberg method.
C
        INTEGER IN, OUT
        REAL SUM, UPPER, LOWER, TOL
        COMMON /INOUT/ IN, OUT
        DATA LOWER/1.0/, UPPER/9.0/, TOL/1.0E-5/
C
C -- f(X) = 1 / X, be careful of X = 0.
C
        F(X) = 1.0 / X
C
        IN = 1
        OUT = 1
        WRITE(OUT, 101)
        CALL ROMBER(LOWER, UPPER, TOL, SUM, F)
        WRITE(OUT, 104) SUM
        STOP
101     FORMAT(/' Romberg integration'/)
104     FORMAT(/' Area =', F10.5/)
        END
        SUBROUTINE ROMBER(LOWER, UPPER, TOL, ANS, F)
C
C -- Numerical integration by the Romberg method.
C
        INTEGER IN, OUT, PIECES, I, P2, NX(16)
        INTEGER NT, II, N, NN, L, NTRA, K, M, J, L2
        REAL X, DELTA, LOWER, UPPER, SUM, TOL
        REAL C, FOTOM, T(136), ANS
        COMMON /INOUT/ IN, OUT
C
        PIECES  = 1
        NX(1) = 1
        DELTA = (UPPER - LOWER) / PIECES
        C = (F(LOWER) + F(UPPER)) / 2.0
        SUM = C
        T(1) = DELTA * C
        N = 1
        NN = 2
5       N = N + 1
          FOTOM = 4.0
          NX(N) = NN
          PIECES = PIECES * 2
          L = PIECES - 1
          L2 = (L+1) / 2
          DELTA = (UPPER - LOWER) / PIECES
C -- Compute trapezoidal sum for 2**(N-1)+1 points.
          DO 10 II = 1, L2
            I = II * 2 - 1
            X = LOWER + DELTA * I
            SUM = SUM + F(X)
10        CONTINUE
```

Figure 9.14: Integration by the Romberg Method

```
            T(NN) = SUM * DELTA
            NTRA = NX(N-1)
            K = N-1
C -- Compute n-th row of T array.
            DO 20 M = 1, K
              J = NN + M
              NT = NX(N-1) + M - 1
              T(J) = (FOTOM * T(J-1) - T(NT)) / (FOTOM - 1)
              FOTOM = FOTOM * 4
20          CONTINUE
            WRITE(OUT, 101) PIECES, T(NN), J, T(J)
            IF (N .LT. 5)  GOTO 40
              IF (T(NN+1) .EQ. 0.0) GOTO 40
                IF (ABS(T(NTRA+1)-T(NN+1)) .LE. ABS(T(NN+1)*TOL)) GOTO 99
                IF (ABS(T(NN-1) - T(J)) .LE. ABS(T(J)*TOL)) GOTO 99
                IF (N .GT. 15) GOTO 100
40          NN = J + 1
         GOTO 5
99       ANS = T(J)
         RETURN
100      WRITE(OUT, 999)
         RETURN
101      FORMAT(1X, I7, F9.5, I5, F9.5)
999      FORMAT(/' Romberg error')
         END
```

Figure 9.14: Integration by the Romberg Method (cont.)

```
        Romberg integration

            2   3.02222      3   2.54815
            4   2.46349      6   2.25919
            8   2.27341     10   2.20472
           16   2.21733     15   2.19773
           32   2.20234     21   2.19724
           64   2.19851     28   2.19723

        Area =   2.19723
```

Figure 9.15: Output: Integration by the Romberg Method

By comparing the values in Figure 9.15 to those in the previous tables, we can see that the Romberg method converges even more rapidly (in this example) than Simpson's method.

Second Run of the Romberg Method—An Exponential Function

Alter the Romberg method so that it calculates the integral:

$$\int_0^1 e^{-x^2}\, dx$$

Change the function statement in the main program so that it reads:

F(X) = EXP(−X∗X)

and change the limits in the data statement of the main program to read:

DATA LOWER/0.0/, UPPER/1.0/

Run the Romberg method with the new formula. The results should look like Figure 9.16. Notice that in this example, convergence is not as rapid as it was with Simpson's method.

```
        Romberg integration

         2   .73137      3   .74718
         4   .74298      6   .74683
         8   .74587     10   .74682
        16   .74658     15   .74682

    Area =    .74682
```

Figure 9.16: Output: Integration of e^{-x^2} *by Romberg's Method*

Third Run of the Romberg Method—A Periodic Function

We will now return to the integral:

$$\int_0^{4\pi} \sin^2 x \, dx$$

that we considered earlier. If we attempt to solve this equation using the Romberg method, we will essentially obtain a value of zero for the first two approximations. However, if we are careful to use a relative tolerance rather than an absolute tolerance to determine convergence, we will find the correct solution after several more iterations.

Make a copy of the Romberg program. Then change the three places that were changed in the Simpson method solution of this problem. The value of π is defined in a data statement:

DATA LOWER/0.0/, PI/3.14159/

and function F(X) is altered to read:

F(X) = SIN(X)∗∗2

The upper limit is defined in the main program:

UPPER = 4.0 ∗ PI

Run the program and compare the output to Figure 9.17.

```
Romberg integration

         2    .00000      3    .00000
         4    .00000      6    .00000
         8   6.28318     10   9.07792
        16   6.28319     15   6.08755
        32   6.28319     21   6.28633

    Area =   6.28633
```

Figure 9.17: Output: Romberg Solution of $\sin^2 x$

Finally, in the next section, we will expand our Romberg integration program to deal with a special case—a function that approaches infinity at one limit to the area beneath the curve. To solve this problem we will develop a technique of adjusting the width of the panels as they approach the infinite limit until we have arrived at a sufficiently accurate value for the total area.

FUNCTIONS THAT BECOME INFINITE AT ONE LIMIT

For each of the above methods, we have used a uniform panel width throughout the desired interval. But it should be apparent that convergence is more difficult (that is, more panels are required) when the function has a greater slope. Conversely, wider panels can be used when the slope is smaller. Furthermore, a function may be infinite at one limit even though the area for the interval is finite.

Consider, for example, the integral of the reciprocal of the square root of x over the range from zero to unity:

$$\int_0^1 \frac{dx}{\sqrt{x}}$$

The exact value of the integral is found to be 2.0 by direct integration.

If we attempt to solve this problem using one of the previous integration programs, we immediately run into a problem. The value of $f(a)$ at the left edge is infinite and so we must choose some other lower limit. If we choose a small lower limit, such as 10^{-11}, convergence will take a very long time. But if we choose a larger value, such as 0.01, for the left limit, the result will be inaccurate.

FORTRAN PROGRAM: ADJUSTABLE PANELS FOR AN INFINITE FUNCTION

One method of dealing with this situation is to start the integration at a value somewhat larger than zero. We will initially choose the left boundary to be a value of 0.1. The integral is then evaluated over the range 0.1 to 1.0. To test the reasonableness of this, we should then evaluate the next region to the left, from 0.01 to 0.1. If the area of this region is also significant, then we must add it to the area obtained for the first region. We then take the next region, from 0.001 to 0.01, and see how large it is. In this way, we can take regions closer and closer to zero, observing the additional area as we go. The program shown in Figure 9.18 uses this approach with the Romberg method.

Running the Program

The program is similar to the previous one, and so it may be best to alter a copy of Figure 9.14. Compile the program, include a copy of subroutine ROMBER, and execute the combination. The statement in ROMBER that prints intermediate values:

C WRITE(OUT, 101) PIECES, T(NN), J, T(J)

should be disabled so that only the final value of each major area will be displayed.

```
C         PROGRAM ROMBM
C
C -- Numerical integration by the Romberg method.
C
          INTEGER IN, OUT
          REAL SUM, UPPER, LOWER, TOL, SUMT
          COMMON /INOUT/ IN, OUT
          DATA LOWER/0.1/, UPPER/1.0/, TOL/1.0E-5/
C
          F(X) = 1.0 / SQRT(X)
C
          IN = 1
          OUT = 1
          SUMT = 0.0
          WRITE(OUT, 101)
10        CALL ROMBER(LOWER, UPPER, TOL, SUM, F)
          UPPER = LOWER
          LOWER = 0.1 * UPPER
          SUMT = SUMT + SUM
          WRITE(OUT, 102) SUM, SUMT, LOWER, UPPER
          IF (ABS(SUM) .GT. TOL) GOTO 10
          STOP
101       FORMAT(/' new area    total area    lower  upper  limits')
102       FORMAT(1X, 0PF9.6, F12.5, 1P2E11.1)
          END
```

Figure 9.18: Romberg Integration with Adjustable Panels

The main program repeatedly calls subroutine ROMBER with limits that are closer and closer to zero. When the new area is less than the tolerance, the process is terminated. The results should look like Figure 9.19.

```
   new area       total area     lower   upper   limits
   1.367544        1.36754       1.0E-02    1.0E-01
    .432456        1.80000       1.0E-03    1.0E-02
    .136754        1.93675       1.0E-04    1.0E-03
    .043246        1.98000       1.0E-05    1.0E-04
    .013675        1.99368       1.0E-06    1.0E-05
    .004325        1.99800       1.0E-07    1.0E-06
    .001368        1.99937       1.0E-08    1.0E-07
    .000432        1.99980       1.0E-09    1.0E-08
    .000137        1.99994       1.0E-10    1.0E-09
    .000043        1.99998       1.0E-11    1.0E-10
    .000014        1.99999       1.0E-12    1.0E-11
    .000004        2.00000       1.0E-13    1.0E-12
```

Figure 9.19: Output: Integration of $1/\sqrt{x}$ Near Zero

If the intermediate print statements in the subroutine are not removed, it will be seen that each of the above regions has been divided into 64 panels. It should be realized that the overall width of each region is one-tenth that of the region immediately to the right and that the final left limit is 10^{-13}. Thus, if the whole region from 10^{-13} to 1.0 were integrated as one unit, it would require over a million million uniformly spaced panels. Such a method would take an extremely long time to obtain the correct answer.

SUMMARY

In this chapter we have developed and compared three different numerical integration programs. The first two—the trapezoidal rule and the Simpson method implementations—use "end correction" for refinement of the final approximation of the area under the curve. The third, more sophisticated method, Romberg's integration, uses a matrix of progressive interpolations, and does not require error correction. We tried these methods on several different functions; in addition, we used a refined Romberg integration program to compute the area under the curve of an infinite function.

All the above integrations except the last were performed on analytic functions with uniformly spaced panels. Other techniques are required with discrete data. One approach is to fit the data to a polynomial function with one of the techniques we have discussed in earlier chapters, and then integrate the resulting polynomial.

EXERCISES

9-1: *Find the value of the integral:*

$$\int_0^{10} x^3 e^{-x}\, dx$$

by the trapezoidal rule, Simpson's rule, and the Romberg method. Which method seems best?

Answer: 5.938; Romberg

9-2: *Find the value of the integral:*

$$\int_0^{10} \frac{x\,dx}{1 + e^x}$$

by the trapezoidal rule, Simpson's rule, and the Romberg method. Which method seems best?

Answer: 0.8220; Simpson

9-3: *Find the value of the integral:*

$$\int_0^{10} \frac{dx}{e^x + e^{-x}}$$

by the trapezoidal rule, Simpson's rule, and the Romberg method. Which method seems best?

Answer: 0.785; trapezoid

9-4: *Find the value of the integral:*

$$\int_0^{10} \frac{dx}{(1 + x^2)^2}$$

by the trapezoidal rule, Simpson's rule, and the Romberg method. Which method seems best?

Answer: 0.785; trapezoid

10

Nonlinear Curve-Fitting Equations

IN CHAPTERS 5 AND 7 we developed computer programs for finding the coefficients to various curve-fitting equations. Since we chose approximating functions with linear coefficients, the resulting equations were linear, and therefore easily solved. Sometimes, however, it is necessary to choose approximating functions with nonlinear coefficients. In this case, the calculation of the coefficients may be more difficult.

In this chapter, we will consider two different techniques for nonlinear curve fitting. In one method, we will linearize the approximating function, and then find the solution to the linear form. Using this first method, we will find curve fits for two different sets of data. First, we will fit a linearized form of the so-called *rational function* to data representing the Clausing factor. Then, we will fit a linearized exponential function to data representing the diffusion of zinc in copper over a given temperature range.

Our second approach to nonlinear curve fitting will be more direct. We will see that there is no general technique for handling approximating functions that have nonlinear coefficients; however, we will investigate a specific method for exponential equations. This method involves eliminating one coefficient and then solving the resulting equations with Newton's method, which we studied in Chapter 8.

Let us begin, then, with our first example of the linearization method.

LINEARIZING THE RATIONAL FUNCTION

A commonly used, nonlinear approximating function is known as the *rational* function. This expression is formed from the ratio of two polynomials:

$$y = \frac{A_1 + A_3x + A_5x^2...}{1 + A_2x + A_4x^2 + A_6x^3...}$$

In this expression, x is the independent variable, y is the dependent variable and A_1, A_2, etc., are the coefficients as usual.

The rational function is nonlinear. However, it can be linearized by the following operations. Both sides of the equation are multiplied by the denominator polynomial, to give:

$$y(1 + A_2x + A_4x^2 + A_6x^3 ...) = A_1 + A_3x + A_5x^2 ...$$

The terms of the new equation can be rearranged to give:

$$y = A_1 - A_2xy + A_3x - A_4x^2y + A_5x^2 - A_6x^3y ...$$

Some of the terms on the right contain the dependent variable, y, as well as the independent variable, x. But remember, both x and y are arrays of known values. It is the unknown coefficients A_1, A_2, ..., A_n that we want. All these coefficients are now linear, and so they can be determined by methods developed in Chapters 5 and 7.

FORTRAN PROGRAM: THE CLAUSING FACTOR FITTED TO THE RATIONAL FUNCTION

A program for producing a least-squares fit to the linearized form of the rational function is given in Figure 10.1. The program can be derived from Figure 7.3 in Chapter 7. Alterations need to be made to the main program and the subroutines INPUT, OUTPUT, and LINFIT. Pay particular attention to the dimension statements.

The data in this program represent the *Clausing factor* as a function of length-to-radius (*L/r*) for cylindrical orifices. When molecules with a long mean free path effuse through a cylindrical orifice, some of the molecules strike the orifice walls and are returned in the opposite direction. The remaining molecules continue through to the other side of the orifice. The Clausing factor gives the fraction of those molecules entering one end of a cylindrical orifice that actually emerge from the other end. The Clausing factor ranges from zero to unity. Of course, the Clausing factor becomes smaller as the cylinder increases in length or decreases in radius. Thus, the ratio of length-to-radius, *L/r*, would appear in a formula for the direct calculation of the Clausing factor.

```
C       PROGRAM CLAUS
C
C -- Least-squares fit to ratio of two polynomials.
C -- Subroutines GAUSSJ, SQUARE, and PLOT required.
C -- May 22, 81
C
        INTEGER IN, OUT, MAXR, MAXC, LINES, NROW, NCOL
        REAL X(20), Y(20), YCALC(20), COEF(4), CORREL
        REAL SIG(4), RESID(20)
        COMMON /INOUT/ IN, OUT, MAXR, MAXC
C
        IN = 1
        OUT = 1
        MAXR = 20
        MAXC = 4
        CALL INPUT(X, Y, NROW)
        CALL LINFIT(X, Y, YCALC, RESID, COEF, SIG,
     *  NROW, NCOL, CORREL)
        CALL OUTPUT(X, Y, YCALC, RESID, COEF, SIG,
     *  NROW, NCOL, CORREL)
        LINES = 2 * (NROW - 1) + 1
        CALL PLOT(X, Y, YCALC, NROW, OUT, LINES)
        STOP
        END
        SUBROUTINE INPUT(X, Y, NROW)
C -- Get values for NROW and arrays X and Y.
C
        INTEGER NROW, I
        REAL X(1), Y(1)
C
        NROW = 10
        X(1) =  0.1
        X(2) =  0.2
        X(3) =  0.5
        X(4) =  1.0
        X(5) =  1.2
        X(6) =  1.5
        X(7) =  2.0
        X(8) =  3.0
        X(9) =  4.0
        X(10) = 6.0
        Y(1) =  0.9524
        Y(2) =  0.9092
        Y(3) =  0.8013
        Y(4) =  0.6720
        Y(5) =  0.6322
        Y(6) =  0.5815
        Y(7) =  0.5142
        Y(8) =  0.4201
        Y(9) =  0.3566
        Y(10) = 0.2755
        RETURN
        END
```

Figure 10.1: The Clausing Factor Fitted to the Ratio of Two Polynomials

```
          SUBROUTINE OUTPUT(X, Y, YCALC, RESID, COEF, SIG,
      *  NROW, NCOL, CORREL)
C
C -- Print out the answers.
C
          INTEGER IN, OUT, NROW, NCOL, I
          REAL X(1), Y(1), YCALC(1), COEF(1), CORREL
          REAL RESID(1), SIG(1)
          COMMON /INOUT/ IN, OUT
C
          WRITE(OUT, 101)
          WRITE(OUT, 102) (I, X(I), Y(I), YCALC(I),
      *  RESID(I), I = 1, NROW)
          WRITE(OUT, 103)
          WRITE(OUT, 106) COEF(1), SIG(1)
          WRITE(OUT, 104) (COEF(I), SIG(I), I = 2, NCOL)
          WRITE(OUT, 105) CORREL
          RETURN
101       FORMAT('    I       X        Y       Y Calc      Resid')
102       FORMAT(I4, F8.1, 3F9.4)
103       FORMAT(/' Coefficients     Errors').
104       FORMAT(0PF8.4, 3X, 1PE12.3)
105       FORMAT(/' Correlation coefficient is', F7.4)
106       FORMAT(0PF8.4, 3X, 1PE12.3, '  Constant term')
          END
          SUBROUTINE LINFIT(X, Y, YCALC, RESID, COEF, SIG,
      *  NROW, NCOL, COR)
C
C -- Least squares fit to NROW sets of X-Y points
C -- using Gauss-Jordan elimination.
C -- Subroutines SQUARE and GAUSSJ needed.
C
          LOGICAL ERROR
          INTEGER NROW, NCOL, I, J, MAXR, MAXC, IN, OUT
          INTEGER INDEX(20,3), NVEC
          REAL X(1), Y(1), YCALC(1), COEF(1), RESID(1)
          REAL A(4,4), XMATR(20,4), SIG(1)
          REAL SUMY, SUMY2, XI, YI, YC, RES, COR, SRS, SEE
          COMMON /INOUT/ IN, OUT, MAXR, MAXC
          DATA NVEC/1/
C
          NCOL = 4
          DO 10 I = 1, NROW
            XI = X(I)
            YI = Y(I)
            XMATR(I,1) = 1.0
            XMATR(I,2) = -XI * YI
            XMATR(I,3) = XI
            XMATR(I,4) = -XI * XI * YI
10        CONTINUE
          CALL SQUARE(XMATR, Y, A, COEF, NROW, NCOL, MAXR, MAXC)
          CALL GAUSSJ(A, COEF, INDEX, NCOL, MAXC, NVEC, ERROR, OUT)
```

Figure 10.1: The Clausing Factor Fitted to the Ratio of Two Polynomials (cont.)

```
         SUMY = 0.0
         SUMY2 = 0.0
         SRS = 0.0
         DO 20 I = 1, NROW
            YI = Y(I)
            YC = 0.0
            DO 15 J = 1, NCOL
15             YC = YC + COEF(J) * XMATR(I,J)
            YCALC(I) = YC
            RES = YC - YI
            RESID(I) = RES
            SRS = SRS + RES*RES
            SUMY = SUMY + YI
            SUMY2 = SUMY2 + YI * YI
20       CONTINUE
         COR = SQRT(1.0 - SRS/(SUMY2 - SUMY*SUMY/NROW))
         IF (NROW .EQ. NCOL) SEE = SQRT(SRS)
         IF (NROW .NE. NCOL) SEE = SQRT(SRS / (NROW - NCOL))
         DO 30 I = 1, NCOL
30          SIG(I) = SEE * SQRT(A(I,I))
         RETURN
         END
```

Figure 10.1: The Clausing Factor Fitted to the Ratio of Two Polynomials (cont.)

Running the Program

Type up the program and run it. The matrix is calculated for four terms, corresponding to a first-order numerator and a second-order denominator. Additional terms can be easily added to the approximating function. The values of NROW and NCOL, and the corresponding dimension statement, must be changed to reflect the actual number of terms. In addition, expressions such as:

$$XMATR(I,5) = XMATR(I,3) * XI$$
$$XMATR(I,6) = XMATR(I,4) * XI$$

must be added if additional terms are desired. The results shown in Figure 10.2 correspond to the equation:

$$y = \frac{1.0017 + 0.238x}{1 + 0.7533x + 0.0917x^2}$$

If the data are fitted with a regular polynomial function, rather than the rational function, the resulting fit will be not as good (for the same number of coefficients in the approximating function).

In our next example of the linearization approach, we will examine an exponential equation. Later in this chapter we will fit the same equation using another, more direct approach.

```
      I        X        Y       Y Calc      Resid
      1        .1     .9524     .9529       .0005
      2        .2     .9092     .9090      -.0002
      3        .5     .8013     .8006      -.0007
      4       1.0     .6720     .6719      -.0001
      5       1.2     .6322     .6324       .0002
      6       1.5     .5815     .5818       .0003
      7       2.0     .5142     .5146       .0004
      8       3.0     .4201     .4199      -.0002
      9       4.0     .3566     .3564      -.0002
     10       6.0     .2755     .2756       .0001

   Coefficients        Errors
     1.0017          4.603E-04   Constant term
      .7533          1.370E-02
      .2380          1.219E-02
      .0917          5.141E-03

   Correlation coefficient is 1.0000
```

Figure 10.2:
Output: The Clausing Factor vs L/r Fitted to a Rational Function

LINEARIZING THE EXPONENTIAL EQUATION

A common nonlinear equation has the form:

$$y = Ae^{Bx}$$

where x is the independent variable, y is the dependent variable, and A and B are the desired coefficients. This equation occurs widely throughout science and engineering because it is the solution to a first-order, differential equation.

This equation can be linearized by taking the logarithm. The result is:

$$\ln y = \ln A + Bx$$

In this form, the dependent variable is $\ln y$ and the unknown coefficients are $\ln A$ and B. Since the new coefficients are linear, a least-squares fit can be obtained with any of the programs developed in Chapters 5 and 7.

FORTRAN PROGRAM: AN EXPONENTIAL CURVE FIT FOR THE DIFFUSION OF ZINC IN COPPER

Figure 10.3 gives a program for finding a least-squares fit to the linearized exponential equation. The program can be derived from the previous program in this chapter. Notice that the calculation of the standard error has been removed and the value of SRS is printed instead.

The data embedded in the input routine represent the diffusion of zinc in copper over the temperature range 600 to 900°C. The diffusion equation is:

$$D = D_0 e^{-Q/RT}$$

where D is the diffusion coefficient in cm sq/sec, D_0 is the diffusion constant in the same units, Q is the activation energy in cal/deg mole, R is the gas constant, and T is the temperature in kelvins. The independent variable, x, is the reciprocal of the temperature in kelvins. The dependent variable, y, is the logarithm of the diffusion coefficient.

```
C       PROGRAM LEXP
C
C -- Least-squares fit for the diffusion of Zn in Cu.
C -- Subroutines GAUSSJ, SQUARE, and PLOT required.
C -- May 22, 81
C
        INTEGER IN, OUT, MAXR, MAXC, LINES, NROW, NCOL
        REAL X(20), Y(20), YCALC(20), COEF(4), CORREL
        REAL SIG(4), RESID(20), T(20), D(20)
        COMMON /INOUT/ IN, OUT, MAXR, MAXC
C
        IN = 1
        OUT = 1
        MAXR = 20
        MAXC = 4
        CALL INPUT(X, Y, T, D, NROW)
        CALL LINFIT(X, Y, YCALC, RESID, COEF, SRS,
     *  NROW, NCOL)
        CALL OUTPUT(T, D, YCALC, COEF, SRS, NROW, NCOL)
        LINES = 2*(NROW - 1) + 1
        CALL PLOT(X, Y, Y, -NROW, OUT, LINES)
        STOP
        END
        SUBROUTINE INPUT(X, Y, T, D, NROW)
C -- Get values for NROW and arrays X and Y.
C
        INTEGER NROW, I
        REAL X(1), Y(1), T(1), D(1)
C
        NROW = 7
        D(1) =   1.4E-12
        D(2) =   5.5E-12
        D(3) =   1.8E-11
        D(4) =   6.1E-11
        D(5) =   1.6E-10
        D(6) =   4.4E-10
        D(7) =   1.2E-9
        DO 10 I = 1, NROW
```

Figure 10.3: A Least-Squares Fit to the Linearized Exponential Equation

```
          T(I) =   550 + 50 * I
          X(I) = 1.0 / (T(I) + 273.15)
          Y(I) = ALOG(D(I))
10      CONTINUE
        RETURN
        END
        SUBROUTINE OUTPUT(X, Y, YCALC, COEF, SRS, NROW, NCOL)
C
C -- Print out the answers.
C
        INTEGER IN, OUT, NROW, NCOL, I
        REAL X(1), Y(1), YCALC(1), COEF(1), DO, Q
        COMMON /INOUT/ IN, OUT
        DATA R/ 1.987/
C
        WRITE(OUT, 101)
        WRITE(OUT, 102) (I, X(I), Y(I), YCALC(I), I = 1, NROW)
        WRITE(OUT, 103)
        WRITE(OUT, 106) COEF(1)
        WRITE(OUT, 104) (COEF(I), I = 2, NCOL)
        DO = EXP(COEF(1))
        Q = -R * COEF(2) / 1000
        WRITE(OUT, 107) DO, Q, SRS
        RETURN
101     FORMAT('  I      T C         D            D Calc')
102     FORMAT(I4, 0PF8.1, 1P2E12.3)
103     FORMAT(/' Coefficients')
104     FORMAT(1PE12.3)
106     FORMAT(1PE12.3, '  Constant term')
107     FORMAT(/' DO =', F9.2, ' cm sq/sec.' /
     *    ' Q  =', F9.2, ' kcal/mole' //' SRS = ', F6.2)
        END
        SUBROUTINE LINFIT(X, Y, YCALC, RESID, COEF, SRS,
     *  NROW, NCOL)
C
C -- Least squares fit to NROW sets of X-Y points
C -- using Gauss-Jordan elimination.
C -- Subroutines SQUARE and GAUSSJ needed.
C
        LOGICAL ERROR
        INTEGER NROW, NCOL, I, J, MAXR, MAXC, IN, OUT
        INTEGER INDEX(20,3), NVEC
        REAL X(1), Y(1), YCALC(1), COEF(1), RESID(1)
        REAL A(4,4), XMATR(20,4), SIG(1), A1
        REAL SUMY, SUMY2, XI, YI, YC, RES, COR, SRS, SEE
        COMMON /INOUT/ IN, OUT, MAXR, MAXC
        DATA NVEC/1/
C
        NCOL = 2
        DO 10 I = 1, NROW
          XMATR(I,1) = 1.0
          XMATR(I,2) = X(I)
10      CONTINUE
```

Figure 10.3: A Least-Squares Fit to the Linearized Exponential Equation (cont.)

```
          CALL SQUARE(XMATR, Y, A, COEF, NROW, NCOL, MAXR, MAXC)
          CALL GAUSSJ(A, COEF, INDEX, NCOL, MAXC, NVEC, ERROR, OUT)
          SUMY = 0.0
          SUMY2 = 0.0
          SRS = 0.0
          A1 = EXP(COEF(1))
          DO 20 I = 1, NROW
            YCALC(I) = A1 * EXP(COEF(2) * X(I))
            IF (Y(I) .EQ. 0.0) RES = 1.0
            IF (Y(I) .NE. 0.0) RES = YCALC(I)/Y(I) - 1.0
            RESID(I) = RES
            SRS = SRS + RES*RES
   20     CONTINUE
          RETURN
          END
```

Figure 10.3: A Least-Squares Fit to the Linearized Exponential Equation (cont.)

Running the Program

Type up the program, run it, and compare the results with Figure 10.4. The diffusion constant can be calculated from the antilogarithm (the exponent) of the first coefficient with the following expression:

$$D0 = EXP(COEF(1))$$

The activation energy, Q, can be obtained from the second coefficient and the gas constant by using the expression:

$$Q = -R * COEF(2)$$

where R, the gas constant, is multiplied by the second coefficient, and its sign is changed.

```
          I     T C        D            D Calc
          1    600.0    1.400E-12     1.313E-12
          2    650.0    5.500E-12     5.433E-12
          3    700.0    1.800E-11     1.943E-11
          4    750.0    6.100E-11     6.135E-11
          5    800.0    1.600E-10     1.740E-10
          6    850.0    4.400E-10     4.499E-10
          7    900.0    1.200E-09     1.073E-09

        Coefficients
        -1.136E+00    Constant term
        -2.290E+04

        D0 =        .32 cm sq/sec.
        Q  =      45.50 kcal/mole

         SRS =    7.00
```

Figure 10.4: Output: The Diffusion of Zinc in Copper (Linearized Fit)

In this example, the sum of residuals squared, *SRS,* is given rather than the usual standard error on the coefficients. The nonlinear transform of the approximating function makes these sigmas meaningless. The calculation of this form of *SRS* is discussed further in the next section, where we will develop our second approach to nonlinear curve fitting.

DIRECT SOLUTION OF THE EXPONENTIAL EQUATION

In the previous section, the exponential equation:

$$y = Ae^{Bx}$$

was linearized by taking the logarithm:

$$\ln y = \ln A + Bx$$

A least-squares fit was then made to this linearized form of the equation. But the coefficients that produce the minimum sum of residuals squared to the linearized form will not, in general, produce the minimum *SRS* for the original, unlinearized equation. We should, therefore, consider a direct, least-squares solution to the nonlinear equation.

In this section, we will derive the curve-fitting equations with respect to the original nonlinearized exponential equation. If we approach the solution to the nonlinear equation the way we approached the linear equation, we will obtain the *SRS* for the approximating function. Then the derivative of *SRS* with respect to each coefficient is set to zero. There is one equation for each unknown. Now, however, the resulting equations are nonlinear and therefore cannot generally be solved.

While there is no universal approach to the solution of nonlinear equations, there are several techniques that can be used for special forms. In the case of the exponential function, the solution is relatively easy. We will follow the usual curve-fit algorithm up to the point of taking the derivatives of the *SRS.* Then we will see a way to eliminate one coefficient from the equation, and we will solve the resulting function using Newton's method.

Calculating the SRS

The residuals for the equation:

$$y = Ae^{Bx}$$

could be defined as:

$$r = Ae^{Bx} - y$$

However, if the data are all measured to about the same relative degree of precision, independent of the magnitude, it will be more meaningful to use a relative residual:

$$r = (Ae^{Bx} - y) / y$$

or

$$r = (Ae^{Bx} / y) - 1$$

The residuals in the new form are squared, and then summed to form the *SRS*. The derivative of the resulting *SRS* is taken with respect to both A and B, and the resulting equations are set to zero. This approach is the same as before. There are two equations and two unknowns.

$$A\Sigma(e^{2Bx} / y^2) - \Sigma(e^{Bx} / y) = 0$$

and

$$A\Sigma(xe^{2Bx} / y^2) - \Sigma(xe^{Bx} / y) = 0$$

Now, however, the resulting equations are nonlinear and so they cannot be solved with the curve-fitting routine used in the previous section.

Eliminating Coefficient A

Fortunately, the coefficient A is linear in this example, and so it can be separated. The first of the two above equations can be rearranged to give:

$$A = \Sigma(e^{Bx} / y) / \Sigma(e^{2Bx} / y^2)$$

Then, this expression for A is substituted into the second equation to give:

$$\Sigma(e^{Bx} / y) \, \Sigma(xe^{2Bx} / y^2)$$
$$- \Sigma(e^{2Bx} / y^2) \, \Sigma(xe^{Bx} / y) = 0 = f(B)$$

Since we have eliminated A, this equation is only a function of B. In Chapter 8 we developed a program to solve nonlinear equations by Newton's method. That approach is applicable here.

Applying Newton's Method

We next take the derivative of the above equation with respect to B. The result is:

$$f'(B) = 2\Sigma(e^{Bx} / y) \, \Sigma(x^2 e^{2Bx} / y^2)$$
$$- \Sigma(xe^{2Bx} / y^2) \, \Sigma(xe^{Bx} / y)$$
$$- \Sigma(e^{2Bx} / y^2) \, \Sigma(x^2 e^{Bx} / y)$$

Before going any further, check to see that all terms in $f(B)$ and $f'(B)$ are homogeneous. The units of each term of $f(B)$ must correspond to:

$$\frac{x}{y^3}$$

and the units of $f'(B)$ must correspond to

$$\frac{x^2}{y^3}$$

FORTRAN PROGRAM: A NONLINEARIZED EXPONENTIAL CURVE FIT

The program shown in Figure 10.5 can be used to find a nonlinear least-squares curve fit to the diffusion equation:

$$D = D_0 e^{-Q/RT}$$

This is the equation we considered in the previous section. Since we are using Newton's method, we need a first approximation for the value of B. We can obtain a good first value from the linear equation for B. But instead of performing the complete linearized fit, we can simply calculate the value of B from Equation 20 of Chapter 5. Notice, however, that the value of y is replaced by ln y.

$$B = \frac{\Sigma[x \ln (y)] - \Sigma(x) \Sigma[\ln (y) / n]}{\Sigma x^2 - (\Sigma x)^2 / n}$$

The instructions for this first approximation to B are included in subroutine LINFIT.

```
C       PROGRAM NLEXP
C
C -- Least-squares fit for the diffusion of Zn in Cu.
C -- Subroutine NEWTON is required.
C -- May 22, 81
C
        INTEGER IN, OUT, MAXR, MAXC, LINES, NROW, NCOL
        REAL YCALC(20), COEF(4), CORREL
        REAL SIG(4), RESID(20), T(20), D(20)
        COMMON /INOUT/ IN, OUT, MAXR, MAXC
        COMMON/FUN/ NROW, A, X(20), Y(20), EX(20)
C
        IN = 1
        OUT = 1
        MAXR = 20
        MAXC = 4
        CALL INPUT(X, Y, T, D, NROW)
        CALL LINFIT(X, Y, YCALC, RESID, COEF, SRS,
     *  NROW, NCOL)
```

Figure 10.5: A Least-Squares Fit to the Nonlinearized Exponential Function

```
        CALL OUTPUT(T, D, YCALC, COEF, SRS, NROW, NCOL)
        STOP
        END
        SUBROUTINE INPUT(X, Y, T, D, NROW)
C -- Get values for NROW and arrays X and Y.
C
        INTEGER NROW, I
        REAL X(1), Y(1), T(1), D(1)
C
        NROW = 7
        D(1) =   1.4E-12
        D(2) =   5.5E-12
        D(3) =   1.8E-11
        D(4) =   6.1E-11
        D(5) =   1.6E-10
        D(6) =   4.4E-10
        D(7) =   1.2E-9
        DO 10 I = 1, NROW
          T(I) =   550 + 50 * I
          X(I) = 1.0 / (T(I) + 273.15)
          Y(I) = D(I)
10      CONTINUE
        RETURN
        END
        SUBROUTINE OUTPUT(X, Y, YCALC, COEF, SRS, NROW, NCOL)
C
C -- Print out the answers.
C
        INTEGER IN, OUT, NROW, NCOL, I
        REAL X(1), Y(1), YCALC(1), COEF(1)
        COMMON /INOUT/ IN, OUT
        DATA R/ 1.987/
C
        WRITE(OUT, 101)
        WRITE(OUT, 102) (I, X(I), Y(I), YCALC(I), I = 1, NROW)
        WRITE(OUT, 103)
        WRITE(OUT, 106) COEF(1)
        WRITE(OUT, 104) (COEF(I), I = 2, NCOL)
        A = COEF(1)
        B = -R * COEF(2) / 1000
        WRITE(OUT, 107) A, B, SRS
        RETURN
101     FORMAT('  I      T C        D           D Calc')
102     FORMAT(I4, 0PF8.1, 1P2E12.3)
103     FORMAT(/' Coefficients')
104     FORMAT(1PE12.3)
106     FORMAT(1PE12.3, '  Constant term')
107     FORMAT(/' DO =', F9.2, ' cm sq/sec.' /
       *  ' Q  =', F9.2, ' kcal/mole' //'  SRS = ', F7.3)
        END
        SUBROUTINE LINFIT(X, Y, YCALC, RESID, COEF, SRS,
       * NROW, NCOL)
```

Figure 10.5: A Least-Squares Fit to the Nonlinearized Exponential Function (cont.)

```
C
C -- Least squares fit to NROW sets of X-Y points.
C -- Subroutine NEWTON needed.
C
      LOGICAL ERROR
      INTEGER NROW, NCOL, I, J, MAXR, MAXC, IN, OUT
      REAL X(1), Y(1), YCALC(1), COEF(1), RESID(1)
      REAL SUMY, SUMY2, XI, YI, YC, RES, COR, SRS, SEE
      EXTERNAL FUNC
      COMMON /INOUT/ IN, OUT, MAXR, MAXC
      COMMON /FUN/ IDUM, A, DUMMY(40), EX(20)
C
      NCOL = 2
      SUMX  = 0.0
      SUMY  = 0.0
      SUMXY = 0.0
      SUMX2 = 0.0
      DO 10 I = 1, NROW
         XI = X(I)
         YI = ALOG(Y(I))
         SUMX  = SUMX  + XI
         SUMY  = SUMY  + YI
         SUMXY = SUMXY + XI*YI
         SUMX2 = SUMX2 + XI*XI
10    CONTINUE
      B = (SUMXY - SUMX*SUMY/NROW) / (SUMX2 - SUMX*SUMX/NROW)
      CALL NEWTON(B, FUNC, ERROR)
      COEF(1) = A
     .COEF(2) = B
      SRS = 0.0
      DO 20 I = 1, NROW
         YCALC(I) = A * EX(I)
         IF (Y(I) .EQ. 0.0) RES = 1.0
         IF (Y(I) .NE. 0.0) RES = YCALC(I)/Y(I) - 1.0
         RESID(I) = RES
         SRS = SRS + RES*RES
20    CONTINUE
      RETURN
      END
      SUBROUTINE FUNC(B, FB, DFB)
C
C -- Calculate function and slope for NEWTON.
C
      COMMON /FUN/ NROW, A, X(20), Y(20), EX(20)
C
      S1 = 0
      S2 = 0
      S3 = 0
      S4 = 0
      S5 = 0
      S6 = 0
      DO 10 I = 1, NROW
```

Figure 10.5: A Least-Squares Fit to the Nonlinearized Exponential Function (cont.)

```
        XI = X(I)
        X2 = XI*XI
        YI = Y(I)
        Y2 = YI*YI
        EX1 = EXP(B * XI)
        EX(I) = EX1
        EX2 = EX1 * EX1
        S1 = S1 + XI * EX2/Y2
        S2 = S2 + EX1/YI
        S3 = S3 + XI * EX1/YI
        S4 = S4 + EX2/Y2
        S5 = S5 + 2 * X2 * EX2/Y2
        S6 = S6 + X2 * EX1/YI
10      CONTINUE
        FB = S1*S2 - S3*S4
        DFB = S2*S5 - S1*S3 - S4*S6
        A = S2/S4
        RETURN
        END
```

Figure 10.5:

A Least-Squares Fit to the Nonlinearized Exponential Function (cont.)

Running the Program

Type up the program shown in the listing. The Newton's method subroutine is the one developed in Chapter 8. The regular WRITE statements have been disabled, however. Run the program and compare the results to Figure 10.6.

```
    I      T C        D           D Calc
    1     600.0    1.400E-12    1.311E-12
    2     650.0    5.500E-12    5.414E-12
    3     700.0    1.800E-11    1.933E-11
    4     750.0    6.100E-11    6.095E-11
    5     800.0    1.600E-10    1.727E-10
    6     850.0    4.400E-10    4.458E-10
    7     900.0    1.200E-09    1.062E-09

 Coefficients
    3.100E-01   Constant term
   -2.287E+04

 DO =        .31  cm sq/sec.
 Q  =      45.44  kcal/mole

    SRS =     .029
```

Figure 10.6: Output: The Diffusion of Zinc in Copper (Nonlinear Fit)

Now, compare the results of Figure 10.4 to Figure 10.6. The diffusion constant D_0 and the activation energy Q are about the same. However, the sum of residuals squared is smaller for the nonlinearized fit. It should be pointed out, however, that if the linearized residuals:

$$r = \ln D_0 - \frac{Q}{RT}$$

are used in *SRS,* the linearized *SRS* will be smaller. Furthermore, both linearized and nonlinearized programs will give exactly the same results if the data precisely follow an exponential equation. Therefore, it might be wise to perform both the linearized and nonlinearized curve fits. If the results are very different, then an error in measuring or in recording the data should be suspected.

SUMMARY

We have seen examples of two approaches to nonlinear curve fitting. The forms of the equations we used in our examples were (1) the rational function, and (2) the exponential function. We used a linearization approach and a direct approach. It is important to keep in mind that neither of the approaches we have studied represents a *general* method for nonlinear curve fitting; rather, we have examined specific techniques that can be used on specific curve-fitting equations.

EXERCISES

10-1: *The band-gap energy,* E_g, *of an intrinsic semiconductor can be determined from the expression:*

$$1/\rho = \sigma = Ae^{-(E_g/2kT)}$$

where ρ is the electrical resistivity, σ is the electrical conductivity, k is Boltzmann's constant, and T is the temperature. The following experimental data were obtained from an intrinsic thermistor:

T C	R ohms
0	1380
4	1200
10	880
18	660
25	488
38	304
43	248
54	170
62	139
77	83

Linearize the equation by taking the logarithm. Then perform a least-squares fit using the experimental data. Since we are only interested in the slope of the resulting fit, we do not have to convert resistance to resistivity. (Of course, the temperature data must be converted to kelvin.) Use a Boltzmann's constant of 8.61×10^{-5} to obtain the band gap energy in electron volts.

Answer: 0.6 ev.

10-2: The time-temperature relationship for the crystallization of a material follows the expression:

$$1/t = Ae^{-Q/RT}$$

where t is the time, Q is the activation energy, R is the gas constant and T is the temperature. Find the activation energy and coefficient A from the given data by performing a least-squares fit on the linearized form of the equation:

$$\ln t = -\ln A + Q/RT$$

Be sure to convert the temperature to kelvin. For the gas constant use $R = 1.987$ cal/deg mole.

T C	t min
350	49
360	40
370	34
380	28
390	24
400	20
410	17
420	14
430	12

Answer: $Q = 15.2$ Kcal/mole, $A = 4470$/min.

11

Advanced Applications: The Normal Curve, the Gaussian Error Function, the Gamma Function, and the Bessel Function

THIS CHAPTER TAKES UP several advanced topics in programming for mathematical applications. The programs use several tools we have developed in this book. We will study three functions, all of which have important applications in mathematics, physics, and engineering: the Gaussian error function, the Gamma function, and the Bessel functions.

To evaluate the Gaussian error function we will write two programs; the first uses Simpson's rule for numeric integration, and the second uses an infinite series expansion. In our discussion of the Gaussian error function we will again take up the topic of diffusion, which we studied in Chapter 10 in the context of nonlinear curve fitting.

We will then examine the Gamma function. Following a study of the special properties of the function, we will develop a program to evaluate it. We will subsequently use this program in the last topic of the chapter—an investigation of numerical solutions to the Bessel equation. We will consider Bessel functions of the first and second kind.

The precision and range of your FORTRAN will be significant issues in running the programs of this chapter. Thus, it will be important to recall the results of the evaluation programs we developed and ran in Chapter 1.

Let us begin by reviewing the concepts of the distribution functions.

THE NORMAL AND CUMULATIVE DISTRIBUTION FUNCTIONS

We saw in Chapter 2 that random errors, introduced during experimental measurement, cause a sequence of observed values to be dispersed about the mean or average. A frequency plot of the resulting data shows a bell-shaped curve. This shape is described as a standard normal distribution or a probability density function, and is illustrated in Figure 11.1.

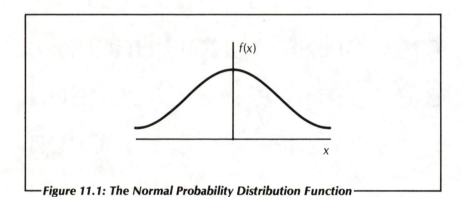

Figure 11.1: The Normal Probability Distribution Function

The normal distribution function is defined by the equation:

$$f(x) = \frac{e^{\frac{-x^2}{2}}}{\sqrt{2\pi}} \qquad (1)$$

This function has a peak, or mean value, at $x = 0$, and ranges from minus infinity to plus infinity. The entire area under the probability-density curve (above the x-axis) is normalized to a value of unity. That is,

$$\int_{-\infty}^{\infty} f(x)dx = 1 \qquad (2)$$

From the symmetry of the curve, it can be seen that the area from $x = 0$ to infinity, the right half of the curve, is equal to half the total area.

The area from minus infinity to b is called the *cumulative distribution function F(x)*.

$$F(x) = \int_{-\infty}^{b} f(x)dx \qquad (3)$$

This integral cannot be solved in closed form, but it is tabulated in hand-books. Sometimes the integral:

$$G(x) = \int_0^b f(x)dx \qquad (4)$$

is given instead. Because the curve is normalized, either integral can readily be obtained from the other. The relationship is:

$$F(x) = G(x) + 0.5, \qquad x \geqslant 0$$

In the next section we will consider some numerical methods for obtaining the area $G(x)$. But first, let us further explore the normal curve.

The Standard Deviation

The area under the normal curve of the measured values is related to the standard deviation. A small standard deviation corresponds to a close grouping of the measured values about the mean. Conversely, a large standard deviation corresponds to values that are spread further from the mean. Two normal distributions are shown in Figure 11.2. Both have the same mean value, but they have different standard deviations. The curve with the smaller standard deviation has the higher peak at zero. However, both have the same area underneath the curve.

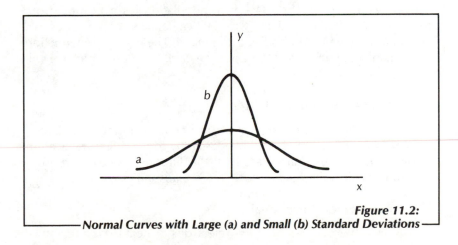

Figure 11.2:
Normal Curves with Large (a) and Small (b) Standard Deviations

Figure 11.3 shows a plot of the distribution function described by Equation 3. This function has a value of 0.5 at $x = 0$ and rises asymptotically to a value of unity as x approaches infinity.

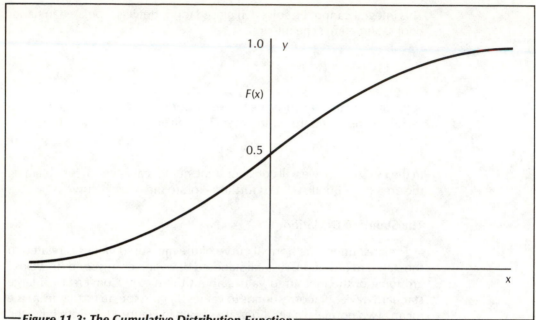

Figure 11.3: The Cumulative Distribution Function

The function $G(x)$ can be used to find the relationship between the standard deviation and the corresponding fraction of a particular sample. For example, $G(1)$ has a value of 0.34. Thus, 34% of the population lies in the range from zero to one standard deviation. Twice that value, 68%, corresponds to a range of one sigma on both sides of the mean. The distribution function can be readily obtained from the Gaussian error function, which we will discuss in the next section.

THE GAUSSIAN ERROR FUNCTION

Before formulating the Gaussian error function and implementing a program to evaluate it, we will digress slightly to a familiar application area—*diffusion*. The equation describing the diffusion of one kind of atom into another will finally lead us back to our Gaussian error function.

Diffusion in a One-Dimensional Slab

Diffusion is the net flow of atoms, electrons or heat from a region of higher concentration to a region of lower concentration. The resulting flux, J, can be described by *Fick's first law*:

$$J = -D \frac{dC}{dx}$$

where C is the concentration, and x is the distance. The *diffusion coefficient* or *diffusivity, D,* is the same quantity we used to describe the diffusion of zinc in copper in Chapter 10. The minus sign reflects the fact that the flux occurs in a direction that is opposite to the concentration gradient. If we are concerned with the flow of atoms, then J is expressed in units of atoms per unit area-sec. The concentration, C, is given in units of atoms per unit volume.

Fick's first law can also be used to describe the flow of electrons in a conductor. We then write:

$$J = \sigma \frac{dV}{dx}$$

where J is the current density, V is the voltage, and x is the distance. Sigma is the electrical conductivity.

In a similar way, the flow of heat can be expressed as:

$$J = k \frac{dT}{dx}$$

where k is the thermal conductivity in units of energy/ area-sec, and T is the temperature.

Fick's second law:

$$\frac{\partial C}{\partial t} = D \frac{\partial^2 C}{\partial x^2}$$

can be used to describe the concentration as a function of position and time. There are many possible solutions to Fick's second law, depending on the boundary conditions. For example, steady-state conditions occur when the concentration no longer changes with time. Since

$$\frac{\partial C}{\partial t} = 0$$

then

$$\frac{D \partial^2 C}{\partial x^2} = 0$$

This implies that a concentration gradient, $\partial C / \partial X$, is uncurving or straight.

Another useful solution to Fick's second law describes the diffusion of one kind of atom into another. The surface concentration of the diffusing species is kept constant for all time. The other material is in the shape of a one-dimensional, semi-infinite slab. For this configuration, the solution to Fick's law is:

$$\frac{C_x - C_0}{C_s - C_0} = 1 - \text{erf}(y) = 1 - \text{erf}\left(\frac{x}{\sqrt{2\,Dt}}\right)$$

In this expression, C_x is the concentration of the diffusing species at time t and a distance x from the surface. C_s is the surface concentration that is constant for all time, and C_0 is the initial, uniform concentration for all x at time equal to zero. D is the diffusion coefficient and y has the value:

$$\frac{x}{2\sqrt{Dt}}$$

If the initial concentration, C_0, is zero, then the equation reduces to:

$$\frac{C_x}{C_s} = 1 - \text{erf}(y) = 1 - \text{erf}\left(\frac{x}{2\sqrt{Dt}}\right)$$

The quantity erf is the Gaussian error function. It is defined as:

$$\text{erf}(y) = \frac{2}{\sqrt{\pi}} \int_0^y e^{-t^2} dt \tag{5}$$

The functions $F(x)$, given in Equation 3, and $G(x)$, given in Equation 4, can be obtained from the error function by the relationship:

$$F(x) = \frac{1 + \text{erf}\left(\frac{x}{\sqrt{2}}\right)}{2}$$

$$G(x) = \frac{\text{erf}\left(\frac{x}{\sqrt{2}}\right)}{2}$$

For example, the range represented by two standard deviations on either side of the mean can be found by the error function:

$$2G(2) = \text{erf}\left(\frac{2}{\sqrt{2}}\right) = 0.954 = 95\%$$

FORTRAN PROGRAM: EVALUATING THE GAUSSIAN ERROR FUNCTION USING SIMPSON'S RULE

The error function cannot be solved in closed form. However, it is possible to obtain particular solutions. A straightforward approach is to use a numerical integration technique such as Simpson's rule. Figure 11.4 gives a program for finding the error function in this way. It can be derived from Figure 9.8. The WRITE statements in subroutine SIMPSON have been disabled so that intermediate results are not displayed.

```
C          PROGRAM ERFSIM
C
C -- Numerical integration by Simpson's rule.
C
           INTEGER IN, OUT
           REAL SUM, UPPER, LOWER, TOL
           COMMON /INOUT/ IN, OUT
           DATA LOWER/0.0/, UPPER/9.0/, TOL/1.0E-5/
           DATA PI/ 3.141592/
C
C -- f(X) = 1 / X, be careful of X = 0.
C
           F(X) = EXP(-X*X)
C
           IN = 1
           OUT = 1

           TWOPI = 2.0 / SQRT(PI)
10         WRITE(OUT, 101)
           READ(IN, 102) UPPER
           IF (UPPER .LT. 0.0) GOTO 99
           SUM = 0.0
           IF (UPPER .GT. 0.0)
     *        CALL SIMPS(LOWER, UPPER, TOL, SUM, F)
           SUM = SUM * TWOPI
           WRITE(OUT, 104) SUM
           GOTO 10
99         STOP
101        FORMAT(/' Arg? ')
102        FORMAT(E10.0)
104        FORMAT(' Erf =', F10.5)
           END
           SUBROUTINE SIMPS(LOWER, UPPER, TOL, SUM, F)
C
C -- Numerical integration by Simpson's rule.
C
           INTEGER IN, OUT, PIECES, I, P2
           REAL X, DELTA, LOWER, UPPER, SUM, TOL
           REAL ENDSUM, ODDSUM, SUM1, EVSUM
           COMMON /INOUT/ IN, OUT
C
           PIECES  = 2
           DELTA = (UPPER - LOWER) / PIECES
           ODDSUM = F(LOWER + DELTA)
           EVSUM = 0.0
           ENDSUM = F(LOWER) + F(UPPER)
           SUM = (ENDSUM + 4 * ODDSUM) * DELTA / 3.0
C          WRITE(OUT, 101) PIECES, SUM
5          PIECES = PIECES * 2
              P2 = PIECES / 2
              SUM1 = SUM
              DELTA = (UPPER - LOWER) / PIECES
              EVSUM = EVSUM + ODDSUM
```

Figure 11.4: The Gaussian Error Function by Simpson's Rule

```
        ODDSUM = 0.0
        DO 10 I = 1, P2
          X = LOWER + DELTA *(2 * I - 1)
          ODDSUM = ODDSUM + F(X)
10      CONTINUE
        SUM = (ENDSUM + 4.0 * ODDSUM + 2.0 * EVSUM)
     *      * DELTA / 3.0
C       WRITE(OUT, 101) PIECES, SUM
        IF (ABS(SUM - SUM1) .GT. ABS(TOL * SUM)) GOTO 5
        RETURN
C 101   FORMAT(1X, I7, F9.5)
        END
```

Figure 11.4: The Gaussian Error Function by Simpson's Rule (cont.)

Running the Program

The program repeatedly cycles, asking the user for input. The program can be terminated by entering a negative value. There are several disadvantages to calculating the error function by this method. The execution time increases and the accuracy decreases as the argument increases. For a six- or seven-digit, floating-point package, the useful range of arguments is from zero to 3.

FORTRAN PROGRAM: EVALUATING THE GAUSSIAN ERROR FUNCTION USING AN INFINITE SERIES EXPANSION

Another way to evaluate the error function is to substitute an infinite series. The new expression is then integrated term by term to produce another infinite series. The result is:

$$\text{erf}(y) = \frac{2}{\sqrt{\pi}} \; e^{-y^2} \sum_{n=0}^{\infty} \frac{2^n y^{2n+1}}{1 \cdot 3 \cdot \ldots \cdot (2n+1)}$$

Figure 11.5 gives a program for evaluating the error function in this way. Type up the program and run it. The user is asked to input an argument to the error function. Then the argument and the resulting function are printed.

The infinite series is evaluated in a loop in function ERF. Each new term is added to the sum. If a particular term does not change the sum by more than the value of TOL (the tolerance), then the routine is terminated and the current value is returned.

```
C       PROGRAM ERFI
C
C -- Gaussian error function.
C
```

Figure 11.5: An Infinite Series Expansion for the Gaussian Error Function

```
         INTEGER IN, OUT
         REAL X, ANS
C
         IN = 1
         OUT = 1

10       WRITE(OUT, 101)
         READ(IN, 102) X
         IF (X .LT. 0.0) GOTO 99
         ANS = ERF(X)
         WRITE(OUT, 104) ANS
         GOTO 10
99       STOP
101      FORMAT(/' Arg? ')
102      FORMAT(E10.0)
104      FORMAT(' Erf =', F10.5)
         END
         FUNCTION ERF(X)
C
C -- Gaussian error function by infinite series.
C
         INTEGER I
         REAL X, X2, SUM, SUM1, TERM
         DATA TOL/1.0E-5/, SQRTPI/ 1.772454/
C
         ERF = 0.0
         IF (X .EQ. 0.0) GOTO 99
         ERF = 1.0
         IF (X .GT. 4.0) GOTO 99
         X2 = X * X
         SUM = X
         TERM = X
         I = 0
10       I = I + 1
           SUM1 = SUM
           TERM = TERM * X2 / (I + 0.5)
           SUM = TERM + SUM1
           IF (TERM .GE. TOL*SUM) GOTO 10
         ERF = 2 * SUM * EXP(-X2) / SQRTPI
99       RETURN
101      FORMAT(1X, I7, F9.5)
         END
```

Figure 11.5:
An Infinite Series Expansion for the Gaussian Error Function (cont.)

Running the Program

The user must enter an argument that is equal to or larger than zero; the resulting function has a range from zero to unity. The result is zero if the argument is zero, and it approaches unity for arguments above 4. On the other hand, the function is approximately equal to its argument over the range from zero to 0.6. The error function has the same relative shape as the cumulative distribution function given in Figure 11.3. Selected values of the error function are given in Figure 11.6. The program will continually cycle, giving new values for the error function until a negative argument is entered.

y	erf(y)
0.0	0.0
0.1	0.1125
0.2	0.2227
0.3	0.3286
0.4	0.4284
0.5	0.5205
0.7	0.6778
1.0	0.8427
2.0	0.9953

Figure 11.6: The Gaussian Error Function

Suppose that zinc is to be diffused into a bar of copper at 900° C. We can use the error function to determine the concentration of copper as a function of time and the distance from the surface. The diffusion coefficient can be calculated from the diffusion constant and the activation energy. These latter quantities were found from the nonlinear curve fit in Chapter 10:

$$D = 0.31\, e^{\frac{-45,420}{1.987(900+273)}}$$

If the distance is chosen to be 0.01 cm and the time is taken as 13 hours, then the argument y in the error function becomes 0.7. This corresponds to an error function of 0.67 and an *error function complement* of 0.33. This means that the concentration of zinc at a depth of 0.01 cm will be 33% of the surface concentration after 13 hours.

In the next section we will see that roundoff errors make it difficult to find a direct solution of the error function complement. We will examine a program that uses two functions to handle both small and large arguments.

THE COMPLEMENT OF THE ERROR FUNCTION

For the above solution to the diffusion equation, we are actually interested in the complement of the error function rather than the error function itself. The complement is defined as:

$$\text{erfc}(y) = \frac{2}{\sqrt{\pi}} \int_y^\infty e^{-t^2} dt \tag{6}$$

The complement is obtained from the relationship:

$$erfc(y) = 1 - erf(y)$$

But if the complement of the error function is always calculated in this way, there will be large roundoff errors for arguments above 3. As the error function approaches unity, the complement approaches zero. Ultimately, all significant figures are lost. Furthermore, the computation time increases as the argument increases.

Equation 6 cannot be integrated by using the alternatives, the trapezoidal rule or Simpson's method, either. A problem occurs in selecting the upper limit for the integral. A value larger than 8 will produce a floating-point underflow because e^{-64} is so small. Yet the area from this point to infinity is significant and cannot be ignored.

FORTRAN PROGRAM: EVALUATING THE COMPLEMENT OF THE ERROR FUNCTION

One solution to this problem is to incorporate two separate routines, one for small arguments and the other for larger arguments. The program given in Figure 11.7 uses this approach. The infinite-series expansion of erf, given in Figure 11.5, is used for smaller arguments. In addition, there is a second function, ERFC, for larger arguments. The complement of the error function is calculated with an asymptotic expansion. The algorithm becomes more accurate as the argument increases. The equation, expressed as a continued fraction, rather than the usual infinite series, is:

$$erfc(y) = \frac{1/[1 + v/\{1 + 2v/[1 + 3v/(1 + ...)]\}]}{\sqrt{\pi}\ ye^{y^2}}$$

where

$$v = \frac{1}{2y^2}$$

```
C       PROGRAM ERFA
C
C  --  Gaussian error function.
C
        INTEGER IN, OUT
        REAL X, E1, E2
C
        IN = 1
        OUT = 1

10      WRITE(OUT, 101)
        READ(IN, 102) X
```

Figure 11.7: The Error Function and its Complement

```
       IF (X .LT. 0.0) GOTO 99
       IF (X .GE. 1.5) GOTO 20
       E1 = ERF(X)
       E2 = 1.0 - E1
       GOTO 30
20     E2 = ERFC(X)
       E1 = 1.0 - E2
30     WRITE(OUT, 104) E1, E2
       GOTO 10
99     STOP
101    FORMAT(/' Arg? ')
102    FORMAT(E10.0)
104    FORMAT(' Erf =', 0PF10.5, ', Erfc =', 1PE12.4)
       END
       FUNCTION ERF(X)
C
C -- Gaussian error function by infinite series.
C
       INTEGER I
       REAL X, X2, SUM, SUM1, TERM
       DATA TOL/1.0E-5/, SQRTPI/ 1.772454/
C
       ERF = 0.0
       IF (X .EQ. 0.0) GOTO 99
       ERF = 1.0
       IF (X .GT. 4.0) GOTO 99
       X2 = X * X
       SUM = X
       TERM = X
       I = 0
10     I = I + 1
         SUM1 = SUM
         TERM = TERM * X2 / (I + 0.5)
         SUM = TERM + SUM1
         IF (TERM .GE. TOL*SUM) GOTO 10
       ERF = 2 * SUM * EXP(-X2) / SQRTPI
99     RETURN
101    FORMAT(1X, I7, F9.5)
       END
       FUNCTION ERFC(X)
C
C -- Complement of Gaussian error function
C -- by asymptotic expansion.
C
       INTEGER I, J, TERMS
       REAL X, X2, SUM, U, V, SQRTPI
       DATA SQRTPI/ 1.772454/, TERMS/12/
C
       X2 = X * X
       V = 0.5 / X2
       U = 1.0 + V * (TERMS + 1)
       DO 10 J = 1, TERMS
```

Figure 11.7: The Error Function and its Complement (cont.)

```
        I = TERMS - J + 1
        SUM = 1.0 + I * V / U
        U = SUM
 10     CONTINUE
        ERFC = EXP(-X2) / (X * SUM * SQRTPI)
        RETURN
        END
```

Figure 11.7: The Error Function and its Complement (cont.)

Running the Program

Type up the new version and run it. The program will now print the error function and its complement. For arguments that are less than or equal to 1.5, the error function is calculated by the infinite series expansion given in Figure 11.5. The complement is then obtained by subtraction from unity. On the other hand, if the argument is larger than 1.5, the complement is calculated from the asymptotic expansion. The error function is determined by subtraction from unity.

Compare the error function from this version of the program with the data in Figure 11.6. Then try the values given in Figure 11.8 to check the complement of the error function.

y	erfc(y)
1.5	$3.390E-2$
2.0	$4.678E-3$
2.5	$4.070E-4$
3.0	$2.209E-5$
3.5	$7.431E-7$
4.0	$1.542E-8$
4.5	$1.966E-10$

Figure 11.8: The Complement of the Error Function

In the next section we will examine the special properties of the Gamma function, and we will implement a program to evaluate it.

THE GAMMA FUNCTION

Related to the error function is the Gamma function, defined by the integral:

$$\Gamma(n) = \int_0^\infty x^{n-1}e^{-x}dx \tag{7}$$

The Gamma function is important because it is part of the solution to Bessel's equation. In addition, it can be used to calculate factorials because of the recursive relationship:

$$\Gamma(n+1) = n\Gamma(n) \tag{8}$$

Since $\Gamma(1) = 1$, we can see that

$$\Gamma(2) = \Gamma(1) = 1 \quad = 1!$$
$$\Gamma(3) = 2\Gamma(2) = 1\cdot2 \quad = 2!$$
$$\Gamma(4) = 3\Gamma(3) = 1\cdot2\cdot3 = 3!$$
$$\cdots$$
$$\Gamma(n) = (n-1)\Gamma(n-1) = (n-1)!$$

Thus, the general formula for using the Gamma function to calculate factorials is:

$$\Gamma(n+1) = n\Gamma(n) = n!$$

Since the Gamma function is defined for all real arguments greater than zero, the "factorial" of noninteger arguments can be defined as well. In addition, we find that:

$$\Gamma(0.5) = -0.5! = \sqrt{\pi}$$

This will be useful in calculating Bessel functions. The Gamma function is also defined for noninteger negative numbers. Its value, however, is infinite for zero and negative integers. A plot of the Gamma function for real arguments is given in Figure 11.9.

The Gamma function can be calculated using a form of Stirling's approximation:

$$\Gamma(x) = \sqrt{\frac{2\pi}{x}}\ x^x e^y$$

where

$$y = \frac{1}{12x} - \frac{1}{360x^3} - x$$

Since this is an asymptotic series, the relative accuracy improves as the argument increases.

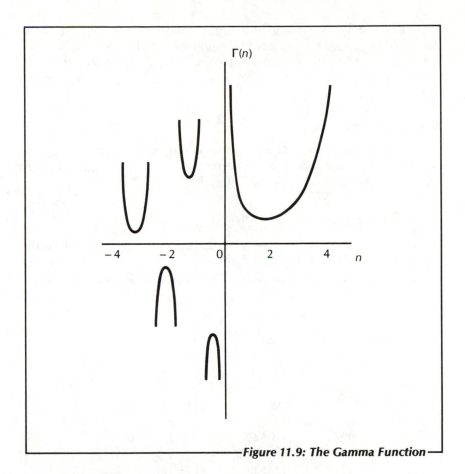

Figure 11.9: The Gamma Function

FORTRAN PROGRAM: EVALUATION OF THE GAMMA FUNCTION

Subroutine GAMMA, given in Figure 11.11, uses Stirling's approximation to calculate the Gamma function. Arguments can be any real number greater than zero and less than about 32. The upper limit depends on the floating-point arithmetic of your FORTRAN. Arguments can also be negative if they are nonintegral. However, the argument cannot be zero or a negative integer.

Positive arguments are incremented by 2, and the Gamma function of the new argument is calculated using a statement function within the function subroutine. The resulting Gamma function is then reduced to the corresponding original argument by using the algorithm:

$$\Gamma(x) = \frac{\Gamma(x+2)}{x(x+1)}$$

This conversion is not needed for larger arguments, but it insures that there will be at least six figures of precision for all values of x.

Negative arguments are incremented until they are positive. The statement function is then called with the new argument. The result is corrected to the original argument.

Running the Program

Type up the program and run it. Try the values given in Figure 11.10 and compare the values for Γ(x) with your results. The program cycles repeatedly until a value of zero is entered.

Since

$$x! = \Gamma(x+1)$$

the program can be readily rewritten so that it will generate factorials rather than the Gamma function.

x	Γ(x)	
1	1	0!
2	1	1!
3	2	2!
4	6	3!
5	24	4!
6	120	5!
0.5	1.7725	$\sqrt{\pi}$
−0.5	−3.5449	Γ(0.5) / (−0.5)
−1.5	2.3633	Γ(−0.5) / (−1.5)
−2.5	0.9453	Γ(−1.5) / (−2.5)

Figure 11.10: Selected Gamma Function Values

```
C       PROGRAM TGAM
C
C   --  The gamma function.
C
        INTEGER IN, OUT
        REAL X, ANS
C
        IN = 1
        OUT = 1

10      WRITE(OUT, 101)
        READ(IN, 102) X
```

Figure 11.11: Evaluation of the Gamma Function

```
            IF (X .EQ. 0.0) GOTO 10
            IF (X .LT. -22.0) GOTO 99
            ANS = GAMMA(X)
            WRITE(OUT, 104) ANS
            GOTO 10
99          STOP
101         FORMAT(/' Arg? ')
102         FORMAT(E10.0)
104         FORMAT(' Gamma =', F11.4)
            END
            FUNCTION GAMMA(X)
C
C -- The gamma function by infinite series.
C
            INTEGER I, J
            REAL X, Y, Z, PI, GAM
            DATA PI/ 3.141593/
C
C -- Function within a function.
C
            G(Y) = SQRT(2 * PI/Y) *
     *      EXP(Y * ALOG(Y) + (1 - 1/(30*Y*Y))/(12*Y)-Y)
     *      / ((Y - 2) * (Y - 1))
C
            IF (X .LT. 0.0) GOTO 10
            GAMMA = G(X + 2.0)
            GOTO 99
C
C -- Increment argument until positive.
C
10          Z = X
            J = -1
            Y = X
20          J = J + 1
            Y = Y + 1.0
            IF (Y .LT. 0.0) GOTO 20
            GAM = G(Y + 2.0)
            DO 30 I = 0, J
              GAM = GAM / (X + I)
30          CONTINUE
            GAMMA = GAM
99          RETURN
            END
```

Figure 11.11: Evaluation of the Gamma Function (cont.)

In the final two sections of this chapter we will introduce the Bessel functions of the first and second kind. *Bessel's equation* has many mathematical and scientific applications; solutions to this differential equation can be found using the Bessel functions. We will examine two FORTRAN implementations of these functions.

BESSEL FUNCTIONS

Bessel's equation:

$$x^2 y'' + xy' + (x^2 - n^2)y = 0$$

arises in the analysis of many different kinds of problems involving circular symmetry. In this equation, x is the independent variable, y is the dependent variable and n is a constant known as the order. This is a nonlinear differential equation that cannot be solved in closed form. One solution to Bessel's equation is:

$$y = J_n(x)$$

where J is the n-order Bessel function of the first kind.

The Bessel functions have been extensively tabulated for particular values of x and n. However, they are difficult to use in this form. For values of x less than about 15, the J Bessel functions can be calculated from the infinite series:

$$J_n(x) = \sum_{k=0}^{n} \frac{(-1)^k}{k!\Gamma(n+k+1)} \left(\frac{x}{2}\right)^{n+2k} \tag{9}$$

On the other hand, the asymptotic expression:

$$J_n(x) = \sqrt{\frac{2}{\pi x}} \cos\left(x - \frac{\pi}{4} - \frac{n\pi}{2}\right) \tag{10}$$

can be used for larger values of x. The Bessel functions J_0 and J_1 are shown in Figure 11.12.

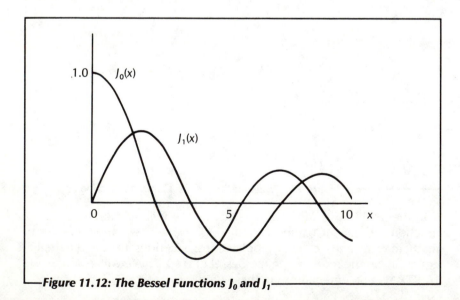

Figure 11.12: The Bessel Functions J_0 and J_1

FORTRAN PROGRAM: BESSEL FUNCTIONS OF THE FIRST KIND

The program shown in Figure 11.14 uses both Equations 9 and 10 to calculate the Bessel functions of the first kind. The order can be zero or a positive number. The argument can also be a noninteger negative number. The Gamma function from the previous section is needed. The infinite series is utilized for arguments less than 15, and the asymptotic expression is used for greater arguments. There may be inaccuracies in this transition region, depending on the floating-point arithmetic of your FORTRAN.

Type up the program and run it. Try the values listed in Figure 11.13 for the order and argument, and compare the results. The program can be terminated by entering an order less than -25.

Order	Argument	Function
1	1	0.4401
0	1	0.7652
1	0.5	0.2423
0	0.5	0.9385
1	10	0.04347
0	10	-0.2459
0	17	-0.1699
10	10	0.2075
0.25	1	0.7522
0.25	1.5	0.6192
0.5	1.5708 $(\pi/2)$	0.6366 $(2/\pi)$
-0.25	1	0.6694
-0.25	1.5	0.3180
-0.75	1.5	-0.2684

Figure 11.13: Selected Bessel Function Values

```
C       PROGRAM BES1
C
C -- Test the Bessel function of the first kind.
C -- Function GAMMA is required.
C
        INTEGER IN, OUT
        REAL X, ANS, ORDER
C
        IN = 1
        OUT = 1

10      WRITE(OUT, 101)
        READ(IN, 102) ORDER
```

Figure 11.14: The Bessel Function of the First Kind

```
         IF (ORDER .LT. -25.0) GOTO 99
         WRITE(OUT, 103)
         READ(IN, 102) X
         ANS = BESJ(X, ORDER)
         WRITE(OUT, 104) ANS
         GOTO 10
99       STOP
101      FORMAT(/' Order? ')
102      FORMAT(E10.0)
103      FORMAT('+ Arg? ')
104      FORMAT('+J Bessel =', F10.4)
         END
         FUNCTION BESJ(X, ORD)
C
C -- Bessel function of the first kind.
C -- Gamma function is required.
C
         INTEGER I
         REAL X, ORD, TOL, PI, TERM, TERM2, SUM, X2
         DATA TOL/1.0E-5/, PI/3.141593/
C
         X2 = X * X
         BESJ = 0.0
         IF ((X .EQ. 0.0) .AND. (ORD .EQ. 1.0)) GOTO 99
         IF (X .GT. 15.0) GOTO 80
         IF (ORD .EQ. 0.0) SUM = 1.0
         IF (ORD .NE. 0.0)
     *   SUM = EXP(ORD * ALOG(X/2)) / GAMMA(ORD + 1)
         TERM2 = SUM
         I = 0
60       I = I + 1
            TERM = TERM2
            TERM2 = -TERM * X2 * 0.25/(I * (ORD + I))
            SUM = SUM + TERM2
         IF (ABS(TERM2) .GT. ABS(SUM * TOL)) GOTO 60
         BESJ = SUM
         GOTO 99
C
C -- Asymptotic expansion for large X.
C
80       BESJ = SQRT(2/(PI*X))*COS(X - PI/4 - ORD*PI/2)
99       RETURN
         END
```

Figure 11.14: The Bessel Functions of the First Kind (cont.)

Since Bessel's equation is a second-order equation, two independent solutions are needed. When the order is not an integer, then both solutions can be obtained from the J Bessel functions. The expression is:

$$y = AJ_n(x) + BJ_{-n}(x)$$

where A and B are constants to be determined from the boundary conditions.

FORTRAN PROGRAM: BESSEL FUNCTIONS OF THE SECOND KIND

If the order of Bessel's equation is an integer, then the solutions $J_n(x)$ and $J_{-n}(x)$ are linearly dependent. In this case, a second, independent solution can be obtained from the expression:

$$y = AJ_n(x) + BY_n(x)$$

where Y is a Bessel function of the second kind.

The program given in Figure 11.17 can be used to calculate Bessel functions of the second kind. While the order can be any real number, it is customary to use these functions only for integer orders. Two different algorithms are utilized. For arguments less than 12, the values of Y_0 and Y_1 are calculated from the expressions:

$$Y_0(x) = \frac{2}{\pi} \sum_{m=0}^{n} (-1)^m \left(\frac{x}{2}\right)^{2m} \frac{[\ln\left(\frac{x}{2}\right) + \gamma - h]}{(m!)^2}$$

and

$$Y_1(x) = -\frac{2}{\pi x} + \frac{2}{\pi} \sum_{m=1}^{n} (-1)^{m+1} \left(\frac{x}{2}\right)^{2m-1} \frac{\left[\ln\left(\frac{x}{2}\right) + \gamma - h + \frac{1}{2m}\right]}{(m!)(m-1)!}$$

where:

$$h = \sum_{r=1}^{m} \frac{1}{r} \quad \text{if} \quad m \geqslant 1$$

and γ is Euler's constant ($\gamma = 0.57721566$). Orders other than 0 and 1 are calculated from the formula:

$$Y_n(x) = \frac{2n}{x} Y_{n-1}(x) - Y_{n-2}(x)$$

For larger arguments, an asymptotic expansion similar to the one used for the J Bessel functions is employed:

$$Y_n(x) = \sqrt{\frac{2}{\pi x}} \sin\left(x - \frac{\pi}{4} - \frac{n\pi}{2}\right)$$

A plot of the functions Y_0 and Y_1 is given in Figure 11.15.

Type up the program and try it out. The argument must be a positive number, since the function goes to minus infinity at zero. Some typical values are shown in Figure 11.16. The program repeatedly cycles. It can be terminated by entering an order that is less than zero.

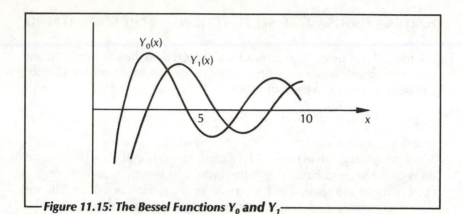

Figure 11.15: The Bessel Functions Y_0 and Y_1

Argument	Y_0	Y_1	Y_2
1	0.088	−0.781	−0.165
2	0.510	−0.107	−0.617
3	0.377	0.325	−0.160
11	−0.169	0.164	0.199
15	0.206	0.021	−0.203

Figure 11.16: Selected Values of the Bessel Function of the Second Kind

```
C         PROGRAM BES2
C
C -- Test the Bessel function of the second kind.
C
          INTEGER IN, OUT
          REAL X, ANS, ORDER
C
          IN = 1
          OUT = 1

10        WRITE(OUT, 101)
          READ(IN, 102) ORDER
          IF (ORDER .LT. 0.0) GOTO 99
20        WRITE(OUT, 103)
          READ(IN, 102) X
          IF (X .LE. 0.0) GOTO 20
          ANS = BESY(X, ORDER)
          WRITE(OUT, 104) ANS
          GOTO 10
99        STOP
```

Figure 11.17: The Bessel Function of the Second Kind

```
101     FORMAT(/' Order? ')
102     FORMAT(E10.0)
103     FORMAT('+ Arg? ')
104     FORMAT('+Y Bessel =', F10.4)
        END
        FUNCTION BESY(X, ORD)
C
C -- Bessel function of the second kind.
C
        INTEGER J
        REAL X, ORD, TOL, PI, TS, TERM2, SUM, X2, PI2, XX
        REAL T, Y0, Y1, YA, YB, YC, SMALL, EULER, FJ1
        DATA PI/3.141593/, PI2/0.6366198/
        DATA SMALL/1.0E-8/, EULER/0.5772157/
C
        IF (X .GE. 12.0) GOTO 80
        XX = 0.5 * X
        X2 = XX * XX
        T = ALOG(XX) + EULER
C
C -- Calculation of Y0.
C
        SUM = 0.0
        TERM = T
        Y0 = T
        J = 0
60      J = J + 1
           IF(J .NE. 1) SUM = SUM + 1.0/(J - 1)
           TS = T - SUM
           TERM = -X2 * TERM / (J*J) * (1.0 - 1.0/(J*TS))
           Y0 = Y0 + TERM
        IF (ABS(TERM) .GE. SMALL) GOTO 60
        Y0 = PI2 * Y0
        IF (ORD .EQ. 0.0) GOTO 90
C
C -- Calculation of Y1 if needed.
C
        TERM = XX * (T - 0.5)
        SUM = 0.0
        Y1 = TERM
        J = 1
70         J = J + 1
           FJ1 = J - 1
           SUM = SUM + 1/(FJ1)
           TS = T - SUM
           T2 = (-X2*TERM) / (J * (FJ1))
           TERM = T2 *((TS - 0.5/J) / (TS + 0.5/(FJ1)))
           Y1 = Y1 + TERM
        IF (ABS(TERM) .GE. SMALL) GOTO 70
        Y1 = PI2 * (Y1 - 1/X)
        IF (ORD .EQ. 1.0) GOTO 95
C
```

Figure 11.17: The Bessel Function of the Second Kind (cont.)

```
C -- Recursive calculation from Y0 and Y1.
C
        TS = 2.0 / X
        YA = Y0
        YB = Y1
        JD = AINT(ORD + 0.01)
        DO 75 J = 2, JD
           YC = TS * (J-1) * YB -YA
           YA = YB
           YB = YC
75      CONTINUE
        BESY = YC
        RETURN
C
C -- Asymptotic expansion for large X.
C
80      BESY = SQRT(2/(PI*X))*SIN(X - PI/4 - ORD*PI/2)
        RETURN
90      BESY = Y0
        RETURN
95      BESY = Y1
        RETURN
        END
```

Figure 11.17: The Bessel Function of the Second Kind (cont)

SUMMARY

In this last chapter we have reviewed several concepts and tools examined in this book. With the tools that we now have available to us we have found that we can implement some rather advanced mathematical applications. We saw FORTRAN programs that evaluate several variations of the Gaussian error function, the Gamma function, and the Bessel functions. In the process of developing these and other programs in this book, we have demonstrated how FORTRAN can be used for technical applications.

EXERCISES

11-1: *Make a copy of the program shown in Figure 11.5 and alter the new version so that it calculates the function G(x) described by Equation 4. Show that G(2) = 0.4772.*

11-2: *Copy Figure 11.11 and alter the new version so that it calculates factorials. Show that the factorial of 0.5 is 0.886.*

11-3: *Write a program to calculate combinations. Start with the program used in the previous problem. The number of ways that six things can be taken two at a time, 6C2, is calculated from the expression:*

$$\frac{6!}{2!(6-2)!}$$

The program should request two numbers: the total number of items and the number of items taken at one time. Show that 6C2 has a value of 15.

Appendix A
Reserved Words and Functions

SOME RESERVED WORDS

The following is a list of FORTRAN keywords (or reserved words) used in this book:

.AND.	CALL	COMMON	COMPLEX	CONTINUE
DATA	DIMENSION	DO	DOUBLE	END
.EQ.	EQUIVALENCE	ERR	EXTERNAL	.FALSE.
FORMAT	FUNCTION	.GE.	GOTO	.GT.
IF	IMPLICIT	INTEGER	.LE.	LOGICAL
.LT.	.NE.	.NOT.	.OR.	PRECISION
PRINT	PROGRAM	READ	REAL	RETURN
STOP	SUBROUTINE	.TRUE.	WRITE	

SOME FORTRAN FUNCTIONS

Care must be taken to insure that the arguments of these functions are of the correct type. For example, SQRT(2.0) not SQRT(2). See Chapter 1 for more details.

Name	Number of Arguments	Type of Arguments	Type of Result	Action
ABS	1	real	real	Absolute value
AINT	1	real	real	Truncate real number
ALOG	1	real	real	Natural logarithm
ATAN	1	real	real	Arctangent
ATAN2	2	real	real	Arctangent arg1/arg2
COS	1	real	real	Cosine
DBLE	1	real	double	Convert single to real

Name	Number of Arguments	Type of Arguments	Type of Result	Action
EXP	1	real	real	Exponent of argument
FLOAT	1	integer	real	Convert integer to real
IFIX	1	real	integer	Convert real to integer
INT	1	real	integer	Convert real to integer
MOD	2	integer	integer	Arg1 modulo arg2
SIGN	2	real	real	Sign of arg1 times arg2
SIN	1	real	real	Sine
SNGL	1	double	real	Convert double to single
SQRT	1	real	real	Square root

FORMAT DESCRIPTORS

Character	Meaning
A	Alphabetic data
D	Double-precision real data
E	Exponential real data
F	Real data without exponent (floating-point)
G	General (or garbage)
H	Literal data (Hollerith)
I	Integer
L	Logical
O	Octal (nonstandard)
P	Scaling factor for real data
R	Right adjust alphabetic (nonstandard)
X	Space

CARRIAGE CONTROL

Printer carriage control is obtained by printing the following characters in column one:

Character	Action (before printing)
1	New page
0	Two lines
blank	One line
+	No advance

Appendix B

Summary of FORTRAN

THE FORTRAN CHARACTER SET

The alphabet	A–Z
The digits	0–9
The special characters	$+ - * / = ( ) . , \$ '$
A space or blank character	

VARIABLE NAMES

A FORTRAN variable name may contain from one to six letters and digits. The first character must be alphabetic. The remaining characters can be alphabetic or numeric.

Variables beginning with the letters I, J, K, L, M, or N default to the type INTEGER. Other variables default to type REAL. This convention may be overridden by specific declaration statements. The other variable types are LOGICAL, DOUBLE PRECISION, and COMPLEX. IMPLICIT typing may also be available.

Example:

```
IMPLICIT DOUBLE PRECISION (P-S)
REAL        NEW, LAST
INTEGER     YESNO, OUT, CMAX
LOGICAL     ERROR, IFLAG
DOUBLE PRECISION SUM1, SUM2
```

ARRAY VARIABLES

FORTRAN arrays may have one, two, or three dimensions. Some implementations allow additional dimensions. The convention for naming array variables is the same as it is for scalar variables. The array size and number of dimensions are declared in a DIMENSION statement. Array elements are referenced by an integer index number, which can run from one to the maximum value declared in the dimension statement. Arrays may also be defined in declaration statements to be REAL, INTEGER, and COMMON.

Examples:

```
DIMENSION X(20), Y(20)
REAL YCALC(20), COEF(3)
INTEGER NX(16)
```

DATA STATEMENTS

Variables are normally preset to zero by the FORTRAN compiler. However, a different value can be selected by using a DATA statement. Of course, the initial value can be changed to something else during execution. Subroutine and function parameters are dummy variables and therefore cannot be initialized with a DATA statement.

Examples:

```
DATA I/1/, J/2/, K/3/
DATA FIRST, MIDDLE, LAST /2HAA/,/2HMM /,/2HZZ /
```

CONSTANTS

FORTRAN constants are numeric or character. Numeric constants can be written as integers or as real numbers. Integers are written as strings of digits. String constants are embedded in apostrophes (single quotes) or can be defined by the letter H. The maximum length of a string constant depends on the word length.

Examples:

```
15   −19253   +7   0   'HOT'   4HCOLD
```

Real numbers are written with a decimal point, a scale factor, or both:

```
379.1275   3.791275E2   3791275E−4
```

The E notation indicates multiplication by powers of 10. Thus, E2 signifies 100.

COMMENTS

Comment lines begin with the letter C. The remainder of the line is ignored by the FORTRAN compiler.

Example:

    C Plot Y and YCALC as a function of X

OPERATIONS

Arithmetic Operators

+	addition
−	subtraction and negation
*	multiplication
/	division
**	exponentiation

Relational Operators

These operations result in a value of .TRUE. or .FALSE. :

.EQ.	equality
.NE.	inequality
.LT.	less than
.GT.	greater than
.LE.	less than or equal to
.GE.	greater than or equal to

Logical Operators

These operations result in a value of .TRUE. or .FALSE. :

    .AND.
    .OR.
    .NOT.

Functional Operators

The built-in functions are listed in Appendix A. Additional functions, called statement functions, can be written as needed. They must be written as single statements.

Example:

    SIZE(X,Y) = SQRT(X*X + Y*Y)

ASSIGNMENT STATEMENTS

Scalar variables and individual array elements are assigned values with the form:

VARIABLE = EXPRESSION

where EXPRESSION is a constant, a variable or a combination of constants and variables resulting in a single value. If the type of EXPRESSION does not match the type of VARIABLE, an automatic type conversion is performed.

THE UNCONDITIONAL BRANCH

The GOTO statement is an unconditional branch. The orderly flow of statement execution is interrupted and the statement at the indicated line number is executed next.

Example:

GOTO 9999

Multiple branch

The statement:

GOTO (300, 400, 500, 600), JK

will cause an unconditional branch to line 300, 400, 500, or 600 if the value of the integer JK is respectively 1, 2, 3, or 4. The value of JK must lie in the range of unity up to the number of statement labels.

CONDITIONAL BRANCHING

The Arithmetic IF

IF (AE) 100, 200, 300

Transfer occurs to statement 100, 200, or 300 if the arithmetic expression AE is respectively less than zero, equal to zero, or greater than zero.

The Logical IF

IF (LE) executable statement

If the logical expression is false, then the remainder of the statement is ignored. On the other hand, if the LE is true, then the statement is executed. The executable statement may not be a DO statement or another logical IF statement.

Examples:

> IF (Y(I) .EQ. 0.0) RES = 1.0
> IF (YESNO .EQ. 'Y') GOTO 10

ITERATIVE STATEMENTS

The DO Loop

> DO 10 I = M1, M2
>
> . . .
>
> 10   CONTINUE

Statements down to 10 are executed with I = M1. At the bottom of the loop, the value of I is incremented. If I is not greater than M2, the loop is executed again.

Example:

> DO 10 I = 1, P2
>   X = LOWER + DELTA *(2 * I − 1)
> 10   CONTINUE

Alternate form:

> DO 10 I = M1, M2, M3
>
> . . .
>
> 10   CONTINUE

With this form, the value of I is increased by M3 after each loop.

INPUT and OUTPUT

Input Data from a Device

> READ (LUN, FNUM) variable or variables list
> READ (LUN, FNUM, END=NUM) variable or variables list

LUN is the logical unit number for the device and FNUM is the format number associated with the input device. The program branches to line number NUM if an end of file occurs.

Examples:

> READ (5, 101) N
> READ (5, 103, END=20) (X(I), I = 1, N)

Output Data to a Device

      WRITE (5, 102) N

      WRITE (5, 104) (X(I), I = 1, N)

SUBROUTINES

A subroutine is called with a statement such as:

      CALL NAME (optional parameter list)

Information can be transferred between the calling program and the subroutine through the parameters or through variables declared in a COMMON statement. The parameters may be variables or the name of another subroutine or function.

      SUBROUTINE NAME (matching dummy parameters)

The first line of the subroutine contains the corresponding dummy parameters.

Bibliography

Daniel, Cuthbert; Wood, Fred; and Gorman, John. *Fitting Equations to Data*. New York: John Wiley & Sons, 1971.

Fike, C. T. *Computer Evaluation of Mathematical Functions*. Englewood Cliffs, N.J.: Prentice-Hall, 1968.

Forsythe, George; Malcolm, Michael; and Moler, Cleve. *Computer Methods for Mathematical Computations.* Englewood Cliffs, N.J.: Prentice-Hall, 1977.

Fox, L., and Mayers, D. F. *Computing Methods for Scientists and Engineers.* New York: Oxford University Press, 1968.

Gilder, Jules. *BASIC Computer Programs in Science and Engineering.* Rochelle Park, N.J.: Hayden, 1980.

Grogono, Peter. *Programming in Pascal*. Revised Edition. Reading, Mass.: Addison-Wesley, 1980.

Hart, John F., et al. *Computer Approximations.* New York: John Wiley & Sons, 1968.

Hastings, Cecil, Jr. *Approximations for Digital Computers.* Princeton, N.J.: University Press, 1955.

Hewlett-Packard. *HP-25 Applications Programs.* Cupertino, Calif., 1975.

Hornbeck, Robert. *Numerical Methods.* New York: Quantum Publishers, 1975.

Hubin, Wilbert. *BASIC Programming for Scientists and Engineers.* Englewood Cliffs, N.J.: Prentice-Hall, 1978.

International Business Machines Corporation. *System/360 Scientific Subroutine Package, Programmer's Manual.* White Plains, N.Y., 1966.

Jennings, Alan. *Matrix Computation for Engineers and Scientists.* New York: John Wiley & Sons, 1977.

Kernighan, Brian W., and Plauger, P. J. *Software Tools.* Reading, Mass.: Addison-Wesley, 1976.

Khabaza, I. M. *Numerical Analysis.* Elmsford, N.Y.: Pergamon Press, 1965.

Kreyszig, Erwin. *Advanced Engineering Mathematics.* New York: John Wiley & Sons, 1967.

Ley, B. James. *Computer Aided Analysis and Design for Electrical Engineers.* New York: Holt, Rinehart and Winston, 1970.

McCormick, John, and Salvadori, Mario. *Numerical Methods in FORTRAN.* Englewood Cliffs, N.J.: Prentice-Hall, 1964.

Miller, Alan R. *BASIC Programs for Scientists and Engineers.* Berkeley, Calif.: Sybex, 1981.

Miller, Alan R. *Pascal Programs for Scientists and Engineers.* Berkeley, Calif.: Sybex, 1981.

Nagin, Paul, and Ledgard, Henry. *BASIC with Style: Programming Proverbs.* Rochelle Park, N.J.: Hayden, 1978.

Ruckdeshel, F. R. *BASIC Scientific Subroutines.* Volume 1. Peterborough, N.H.: Byte/McGraw-Hill, 1981.

Scheaffer, R. L., and Mendenhall, W. *Introduction to Probability: Theory and Applications.* N. Scituate, Mass.: Duxbury Press, 1975.

Smith, J. *Advanced Analysis with the Sharp 5100 Scientific Calculator.* New York: John Wiley & Sons, 1979.

Sokolnikoff, Ivan, and Sokolnikoff, Elizabeth. *Higher Mathematics for Engineers and Physicists.* New York: McGraw-Hill, 1941.

Vandergraft, J. S. *Introduction to Numerical Computations.* New York: Academic Press, 1978.

Walpole, Ronald, and Myers, Raymond. *Probability and Statistics for Engineers and Scientists.* New York: Macmillan Co., 1972.

Wylie, Clarence, Jr. *Advanced Engineering Mathematics.* Second Edition. New York: McGraw-Hill, 1960.

Index

Adjustable panels, 220
Array, 29
  subscript, 30
  index, 30
ASCII bell character, 10
Asymptotic expansion, 253
Average, 13

Back substitution, 57
Bell-shaped curve, 15
Bessel function
  first kind, 260
  second kind, 263
Bessel's equation, 260
Bubble sort, 127

Clausing factor, 230
Coefficient matrix, 49
  determinant of, 49
  inverse of, 63, 75–76, 148
Column index, 33
Column vector, 31
Complement of the error
  function, 252
Complex coefficient, 82
Complex conjugate, 84
Complex number, 82
Conformable matrix, 35
Console bell, 10
Convergence
  lack of, 188
  to a root, 178
Correlation coefficient, 116
Cramer's rule, 49
Cross product, vector, 32
Cumulative distribution
  function, 244
Curve-fitting equation, 111

Data, 146
Determinant, 40, 71
Diffusion, 246
  of zinc in copper, 230
Diffusion coefficient, 231,
  247–48
Diffusivity, 247
Dispersion, 15

Dot product, vector, 32, 37
Dummy array, 18
Dummy parameter, 184
Dummy variable, 18, 116
Dynamic range, 1

End correction, 207, 213
Equation of state, 163
Exponential curve fit, 230
Exponential equation, 234
  linearizing, 230
EXTERNAL statement, 184

Factorials, 256
Failure to converge, 188
Fick's first law, 246
Fick's second law, 247
Floating–point operations
  double precision, 7
  dynamic range of, 1
  precision of, 1
  roundoff error in, 3
  test of, 2
Floating-point package test, 77
Floating-point underflow, 4
Function statement, 31

Gamma function, 255
Gauss elimination, 55
Gauss-Jordan elimination, 63
Gauss-Seidel iterative
  method, 89
Gaussian distribution, 15
Gaussian error function, 246
Gaussian random numbers, 24

Heat capacity, 156
Hilbert matrix, 75

Identity matrix, 34, 43
Ill conditioning, 148, 170
Ill-conditioned matrix, 74
Infinite series expansion, 250
Inverse matrices, 43

Keyboard, entering data from, 9
Kirchhoff voltage law, 51, 83

Least-squares curve-fitting, 99, 139
  criterion for, 111
Linear combination, 68
Linear dependence, 50
Linear equation, 47
Linearly dependent, 263
Local variable, 116
Logical unit numbers, *xiii*
Loop current, 51, 83

Magnitude, 31
Matrix, 33, 146
  addition, 35
  conformable, 35
  division, 43
  inversion, 69
  multiplication, 35, 148
  subtraction, 35
Mean value, 13
Multiple constant vector, 69
Multiple roots, 193

Newton's method, 175
Nonlinear coefficient, 164
Nonlinear curve-fitting, 225
Nonlinear equation, 47, 95
Nonrecursive quick sort, 134
Normal distribution, 15
Normal distribution function, 244
Numerical integration, 199

Parabolic curve fit, 140
Phasor angle, 88
Pivot element, 56
Plotter routine, 103
Point relaxation, 91
Polar magnitude, 88
Polynomial equation, 139
Precision, floating-point, 1
Principal diagonal, 33

Random number generator, 20, 100, 127
  evaluation of, 21
  standard deviation of, 21
Random numbers, using π to produce, 23
Rational function, linearization of, 226
Recursion, 134
Representation, double precision floating-point, 7

Residual, 111
Romberg method, 215
  T matrix in, 216
Roundoff error, 3, 76, 89
Row index, 33
Row interchange, 56, 91
Row vector, 30

Sampling errors, 24
Scalar multiplication, 34
Scalar multiplication of vectors, 31
Scalar variable, 29
Shell-Metzner sort, 133
Simpson's method, 209
Simpson's rule, 248
Simultaneous best fit, 78
Simultaneous equation, 82
Simultaneous solutions, 47
SIN function, test of, 4
Singular matrix, 43, 50, 63, 68, 74
Solution of simultaneous linear equations, 43
SQRT Function, argument of, 8
Square matrix, 33
Standard deviation, 15, 117, 245
  of a random number generator, 21
Standard error, 148
  of the estimate (SEE), 117, 148
  of the coefficients, 118
Stirling's approximation, 256
Swap function, 130
Symmetric, 34

Three-variable equation, 163
Tolerance, 2
Transpose, 34, 146
Trapezoidal rule, 201
Types, improper mixing of, 8

Unit matrix, 34, 63

Vapor pressure, 160, 176
Vapor pressure equation, 194
Vector, 30, 146
  addition, 31
  dot product, 32, 37
  product, 32
  scalar product, 32

Selections from The SYBEX Library

Buyer's Guide

PORTABLE COMPUTERS
by Sheldon Crop and Doug Mosher
128 pp., illustr., Ref. 0-144
"This book provides a clear and concise introduction to the expanding new world of personal computers."—Mark Powelson, Editor, San Francisco Focus Magazine

SELECTING THE RIGHT DATA BASE SOFTWARE
SELECTING THE RIGHT WORD PROCESSING SOFTWARE
SELECTING THE RIGHT SPREADSHEET SOFTWARE
by Kathy McHugh and Veronica Corchado
80 pp., illustr., Ref. 0-174, 0-177, 0-178
This series on selecting the right business software offers the busy professional concise, informative reviews of the best available software packages.

Introduction to Computers

THE COLLEGE STUDENT'S COMPUTER HANDBOOK
by Bryan Pfaffenberger
350 pp., illustr., Ref. 0-170
This friendly guide will aid students in selecting a computer system for college study, managing information in a college course, and writing research papers.

COMPUTER CRAZY
by Daniel Le Noury
100 pp., illustr., Ref. 0-173
No matter how you feel about computers, these cartoons will have you laughing about them.

DON'T!
(or How to Care for Your Computer)
by Rodnay Zaks
214pp., 100 illustr., Ref. 0-065
The correct way to handle and care for all elements of a computer system, including what to do when something doesn't work.

INTERNATIONAL MICROCOMPUTER DICTIONARY
120 pp., Ref. 0-067
All the definitions and acronyms of microcomputer jargon defined in a handy pocket-sized edition. Includes translations of the most popular terms into ten languages.

FROM CHIPS TO SYSTEMS: AN INTRODUCTION TO MICROPROCESSORS
by Rodnay Zaks
552 pp., 400 illustr., Ref. 0-063
A simple and comprehensive introduction to microprocessors from both a hardware and software standpoint: what they are, how they operate, how to assemble them into a complete system.

Personal Computers

IBM

THE ABC'S OF THE IBM® PC
by Joan Lasselle and Carol Ramsay
100 pp., illustr., Ref. 0-102
This is the book that will take you through the first crucial steps in learning to use the IBM PC.

THE BEST OF IBM® PC SOFTWARE
by Stanley R. Trost
144 pp., illustr., Ref. 0-104
Separates the wheat from the chaff in the world of IBM PC software. Tells you what to expect from the best available IBM PC programs.

THE IBM® PC-DOS HANDBOOK
by Richard Allen King
144 pp., illustr., Ref. 0-103
Explains the PC disk operating system, giving the user better control over the system. Get the most out of your PC by adapting its capabilities to your specific needs.

BUSINESS GRAPHICS FOR THE IBM® PC
by Nelson Ford
200 pp., illustr., Ref. 0-124
Ready-to-run programs for creating line graphs, complex illustrative multiple bar graphs, picture graphs, and more. An ideal way to use your PC's business capabilities!

THE IBM® PC CONNECTION
by James W. Coffron
200 pp., illustr., Ref. 0-127
Teaches elementary interfacing and BASIC programming of the IBM PC for connection to external devices and household appliances.

BASIC EXERCISES FOR THE IBM® PERSONAL COMPUTER
by J.P. Lamoitier
252 pp., 90 illustr., Ref. 0-088
Teaches IBM BASIC through actual practice, using graduated exercises drawn from everyday applications.

USEFUL BASIC PROGRAMS FOR THE IBM® PC
by Stanley R. Trost
144 pp., Ref. 0-111
This collection of programs takes full advantage of the interactive capabilities of your IBM Personal Computer. Financial calculations, investment analysis, record keeping, and math practice—made easier on your IBM PC.

YOUR FIRST IBM® PC PROGRAM
by Rodnay Zaks
182 pp., illustr., Ref. 0-171
This well-illustrated book makes programming easy for children and adults.

YOUR IBM® PC JUNIOR
by Douglas Hergert
250 pp., illustr., Ref. 0-179
This comprehensive reference guide to IBM's most economical microcomputer offers many practical applications and all the helpful information you'll need to get started with your IBM PC Junior.

DATA FILE PROGRAMMING ON YOUR IBM® PC
by Alan Simpson
275 pp., illustr., Ref. 0-146
This book provides instructions and examples of managing data files in BASIC. Programming designs and developments are extensively discussed.

Software and Applications

Operating Systems

THE CP/M® HANDBOOK
by Rodnay Zaks
320 pp., 100 illustr., Ref 0-048
An indispensable reference and guide to CP/M—the most widely-used operating system for small computers.

MASTERING CP/M®
by Alan R. Miller
398 pp., illustr., Ref. 0-068
For advanced CP/M users or systems programmers who want maximum use of the CP/M operating system . . . takes up where our *CP/M Handbook* leaves off.

**THE BEST OF
CP/M® SOFTWARE**
by John D. Halamka
250 pp., illustr., Ref. 0-100
This book reviews tried-and-tested, commercially available software for your CP/M system.

REAL WORLD UNIX™
by John D. Halamka
250 pp., illustr., Ref. 0-093
This book is written for the beginning and intermediate UNIX user in a practical, straightforward manner, with specific instructions given for many special applications.

THE CP/M PLUS™ HANDBOOK
by Alan R. Miller
250 pp., illustr., Ref. 0-158
This guide is easy for the beginner to understand, yet contains valuable information for advanced users of CP/M Plus (Version 3).

Business Software

VISICALC® FOR SCIENCE AND ENGINEERING
by Stanley R. Trost and Charles Pomernacki
225 pp., illustr., Ref. 0-096
More than 50 programs for solving technical problems in the science and engineering fields. Applications range from math

Business Applications

COMPUTER POWER FOR YOUR LAW OFFICE
by Daniel Remer
225 pp., Ref. 0-109
How to use computers to reach peak productivity in your law office, simply and inexpensively.

COMPUTER POWER FOR YOUR ACCOUNTING OFFICE
by James Morgan
250 pp., illustr., Ref. 0-164
This book is a convenient source of information about computerizing you accounting office, with an emphasis on hardware and software options.

Languages

C

UNDERSTANDING C
by Bruce Hunter
200 pp., Ref 0-123
Explains how to use the powerful C language for a variety of applications. Some programming experience assumed.

FIFTY C PROGRAMS
by Bruce Hunter
200 pp., illustr., Ref. 0-155
Beginning as well as intermediate C programmers will find this a useful guide to programming techniques and specific applications.

BASIC

YOUR FIRST BASIC PROGRAM
by Rodnay Zaks
150pp. illustr. in color, Ref. 0-129
A "how-to-program" book for the first time computer user, aged 8 to 88.

FIFTY BASIC EXERCISES
by J. P. Lamoitier
232 pp., 90 illustr., Ref. 0-056
Teaches BASIC by actual practice, using graduated exercises drawn from every-day applications. All programs written in Microsoft BASIC.

EXECUTIVE PLANNING WITH BASIC
by X. T. Bui
196 pp., 19 illustr., Ref. 0-083
An important collection of business management decision models in BASIC, including Inventory Management (EOQ), Critical Path Analysis and PERT, Financial Ratio Analysis, Portfolio Management, and much more.

BASIC PROGRAMS FOR SCIENTISTS AND ENGINEERS
by Alan R. Miller
318 pp., 120 illustr., Ref. 0-073
This book from the "Programs for Scientists and Engineers" series provides a library of problem-solving programs while developing proficiency in BASIC.

CELESTIAL BASIC
by Eric Burgess
300 pp., 65 illustr., Ref. 0-087
A collection of BASIC programs that rapidly complete the chores of typical astronomical computations. It's like having a planetarium in your own home! Displays apparent movement of stars, planets and meteor showers.

YOUR SECOND BASIC PROGRAM
by Gary Lippman
250 pp., illustr., Ref. 0-152
A sequel to *Your First BASIC Program*, this book follows the same patient, detailed approach and brings you to the next level of programming skill.

Pascal

INTRODUCTION TO PASCAL (Including UCSD Pascal™)
by Rodnay Zaks
420 pp., 130 illustr., Ref. 0-066
A step-by-step introduction for anyone wanting to learn the Pascal language. Describes UCSD and Standard Pascals. No technical background is assumed.

THE PASCAL HANDBOOK
by Jacques Tiberghien
486 pp., 270 illustr., Ref. 0-053
A dictionary of the Pascal language, defining every reserved word, operator, procedure and function found in all major versions of Pascal.

APPLE® PASCAL GAMES
by Douglas Hergert and Joseph T. Kalash
372 pp., 40 illustr., Ref. 0-074
A collection of the most popular computer games in Pascal, challenging the reader not only to play but to investigate how games are implemented on the computer.

INTRODUCTION TO THE UCSD p-SYSTEM™
by Charles W. Grant and Jon Butah
300 pp., 10 illustr., Ref. 0-061
A simple, clear introduction to the UCSD Pascal Operating System; for beginners through experienced programmers.

PASCAL PROGRAMS FOR SCIENTISTS AND ENGINEERS
by Alan R. Miller
374 pp., 120 illustr., Ref. 0-058
A comprehensive collection of frequently used algorithms for scientific and technical applications, programmed in Pascal. Includes such programs as curve-fitting, integrals and statistical techniques.

DOING BUSINESS WITH PASCAL
by Richard Hergert and Douglas Hergert
371 pp., illustr., Ref. 0-091
Practical tips for using Pascal in business programming. Includes design considerations, language extensions, and applications examples.

Assembly Language Programming

PROGRAMMING THE 6502
by Rodnay Zaks
386 pp., 160 illustr., Ref. 0-046
Assembly language programming for the 6502, from basic concepts to advanced data structures.

6502 APPLICATIONS
by Rodnay Zaks
278 pp., 200 illustr., Ref. 0-015
Real-life application techniques: the input/output book for the 6502.

ADVANCED 6502 PROGRAMMING
by Rodnay Zaks
292 pp., 140 illustr., Ref. 0-089
Third in the 6502 series. Teaches more advanced programming techniques, using games as a framework for learning.

PROGRAMMING THE Z80
by Rodnay Zaks
624 pp., 200 illustr., Ref. 0-069
A complete course in programming the Z80 microprocessor and a thorough introduction to assembly language.

Z80 APPLICATIONS
by James W. Coffron
288 pp., illustr., Ref. 0-094
Covers techniques and applications for using peripheral devices with a Z80 based system.

PROGRAMMING THE 6809
by Rodnay Zaks and William Labiak
362 pp., 150 illustr., Ref. 0-078
This book explains how to program the 6809 in assembly language. No prior programming knowledge required.

PROGRAMMING THE Z8000
by Richard Mateosian
298 pp., 124 illustr., Ref. 0-032
How to program the Z8000 16-bit microprocessor. Includes a description of the architecture and function of the Z8000 and its family of support chips.

PROGRAMMING THE 8086/8088
by James W. Coffron
300 pp., illustr., Ref. 0-120
This book explains how to program the 8086 and 8088 in assembly language. No prior programming knowledge required.

Other Languages

A MICROPROGRAMMED APL IMPLEMENTATION
by Rodnay Zaks
350 pp., Ref. 0-005
An expert-level text presenting the complete conceptual analysis and design of an APL interpreter, and actual listing of the microcode.

Hardware and Peripherals

MICROPROCESSOR INTERFACING TECHNIQUES
by Rodnay Zaks and Austin Lesea
456 pp., 400 illustr., Ref. 0-029
Complete hardware and software interconnect techniques, including D to A conversion, peripherals, standard buses and troubleshooting.

THE RS-232 SOLUTION
by Joe Campbell
225 pp., illustr., Ref. 0-140
Finally, a book that will show you how to correctly interface your computer to any RS-232-C peripheral.

USING CASSETTE RECORDERS WITH COMPUTERS
by James Richard Cook
175 pp., illustr., Ref. 0-169
Whatever your computer or application, you will find this book helpful in explaining details of cassette care and maintenance.

SYBEX Computer Books are different.

Here is why . . .

At SYBEX, each book is designed with you in mind. Every manuscript is carefully selected and supervised by our editors, who are themselves computer experts. We publish the best authors, whose technical expertise is matched by an ability to write clearly and to communicate effectively. Programs are thoroughly tested for accuracy by our technical staff. Our computerized production department goes to great lengths to make sure that each book is well-designed.

In the pursuit of timeliness, SYBEX has achieved many publishing firsts. SYBEX was among the first to integrate personal computers used by authors and staff into the publishing process. SYBEX was the first to publish books on the CP/M operating system, microprocessor interfacing techniques, word processing, and many more topics.

Expertise in computers and dedication to the highest quality product have made SYBEX a world leader in computer book publishing. Translated into fourteen languages, SYBEX books have helped millions of people around the world to get the most from their computers. We hope we have helped you, too.

For a complete catalog of our publications please contact:

U.S.A.
SYBEX, Inc.
2344 Sixth Street
Berkeley,
California 94710
Tel: (800) 227-2346
     (415) 848-8233
Telex: 336311

FRANCE
SYBEX
4 Place Félix-Eboué
75583 Paris Cedex 12
France
Tel: 1/347-30-20
Telex: 211801

GERMANY
SYBEX-VERLAG
Heyestr. 22
4000 Düsseldorf 12
West Germany
Tel: (0211) 287066
Telex: 08 588 163